Globalization in the 21st Century

Globalization in the 21st Century

Labor, Capital, and the State on a World Scale

Edited by

Berch Berberoglu

GLOBALIZATION IN THE 21ST CENTURY
Copyright © Berch Berberoglu, 2010.
Softcover reprint of the hardcover 1st edition 2010 978-0-230-61852-7

First published in 2010 by
PALGRAVE MACMILLAN®
in the United States—a division of St. Martin's Press LLC,
175 Fifth Avenue, New York, NY 10010.

Where this book is distributed in the UK, Europe and the rest of the world,
this is by Palgrave Macmillan, a division of Macmillan Publishers Limited,
registered in England, company number 785998, of Houndmills,
Basingstoke, Hampshire RG21 6XS.

Palgrave Macmillan is the global academic imprint of the above companies
and has companies and representatives throughout the world.

Palgrave® and Macmillan® are registered trademarks in the United States,
the United Kingdom, Europe and other countries.

ISBN 978-1-349-38113-5   ISBN 978-0-230-10639-0 (eBook)
DOI 10.1057/ 9780230106390
Library of Congress Cataloging-in-Publication Data

　　Globalization in the 21st century : labor, capital, and the state on a
world scale / edited by Berch Berberoglu.
　　　p. cm.
　　Includes bibliographical references and index.

　　1. Globalization—Economic aspects. 2. Capitalism. 3. State, The.
I. Berberoglu, Berch. II. Title: Globalization in the twenty first century.

HF1359.G581973 2010
337—dc22   2009039962

A catalogue record of the book is available from the British Library.

Design by Newgen Imaging Systems (P) Ltd., Chennai, India.

First edition: April 2010

10 9 8 7 6 5 4 3 2 1

Contents

Acknowledgments

A project such as this involves the contributions of many people, first and foremost the contributors of this volume: Jan Nedervene Pieterse, Alan Spector, James Petras, Henry Veltmeyer, Johnson Makoba, Alvin So, Lourdes Benería, and Marty Orr. I would like to thank them for providing an excellent set of articles that constitute the ten chapters that make up this book. In addition, I would like to thank friends and colleagues who have over the years contributed to discussions and debates on the nature of the global economy, capitalist development, imperialism and globalization, and struggles against neoliberal globalization (i.e., capitalist imperialism) by mass movements around the world. These include Albert J. Szymanski, James Petras, Larry Reynolds, Blain Stevenson, Walda Katz-Fishman, Richard Dello Buono, David L. Harvey, Johnson W. Makoba, and many others. I thank them all for providing much stimulus through their input to critical discussions that led to the formation of the topics taken up in this book.

An earlier version of the chapters by Alan Spector, James Petras and Henry Veltmeyer, Alvin So, and Marty Orr were originally published as articles in a Special Issue of the journal *International Review of Modern Sociology,* Vol. 33 (2007), on "Globalization" which I edited as Guest Editor. I would like to thank the Editor-in-Chief of this journal, Professor Sunil Kukreja, for allowing me to include revised and updated versions of these articles in this book.

I would also like to thank Patricia Zline of Rowman and Littlefield Publishing Group for giving me permission to reprint a revised version of Chapter 5 of my book *Globalization and Change: The Transformation of Global Capitalism* (Lexington Books, 2005). My thanks also go to Routledge (an imprint of the Taylor & Francis Group) for granting permission to compile excerpts from Lourdes Benería's book *Gender, Development, and Globalization* into a chapter included in this book.

I would like to thank Julia Cohen, my editor at Palgrave Macmillan, for providing me the kind of support and guidance an author and editor could only wish for; I thank her for making the publication process a most rewarding and pleasant experience.

Finally, I would like to thank my wife Suzan for her continued encouragement and support for the completion of this project.

Preface

The global economy is in crisis and globalization is in transition from its neoliberal form propagated by a single capitalist superpower, the United States, to a new regulated global economy with a multipolar base when multiple centers of economic and military power will come to define the nature and dynamics of twenty-first-century globalization.

Neoliberal globalization, which was the driving force of the capitalist expansionary process across the world during the past three decades, came to exemplify the most advanced stage of capitalist imperialism in the late twentieth century. Dominating the economies of less developed capitalist states through this process, it facilitated the transformation of societies along capitalist lines and effected their integration into the global economy to further the process of capital accumulation on a worldwide scale.

The outcome of this process, however, was soon to overwhelm many countries around the globe, when more and more countries were forced to bear the cost of this impact as they witnessed the denationalization of their economies through privatization, transnational corporate control, increasing foreign debt, deteriorating terms of trade, uneven distribution of income and wealth, and increasing class polarization. Thus, beginning in Latin America and spreading to Asia, Eastern Europe, and elsewhere, neoliberal globalization came to thoroughly dominate the global economy.

As the contradictions of neoliberal globalization began to surface and its devastating impact began to be felt by millions across the globe by century's end, neoliberalism came under serious criticism, with mounting opposition from broad segments of the population. Thus, movements across the world coalesced to challenge the foundations of the neoliberal capitalist order by mass protests against the pillars of global capitalism—the World Bank, the International Monetary Fund, the World Trade Organization, and the leading imperialist states—as tools of transnational corporations and global capital that had imposed its rule over the global capitalist system.

It is the protracted struggles of people across the world and their determination to fight against the forces of neoliberal globalization imposed by global capitalism, coinciding with the global economic and financial crisis during the first decade of the twenty-first century, that brought the collapse of the neoliberal economic order and opened the way for a thorough transformation of capitalism on a global scale—one that will lead to the emergence of new forces on the world scene and bring about a new multipolar reality that will shape the nature and future direction of globalization in the twenty-first century.

This project has been in the making over the past several years, and its publication at this critical juncture, when the global economic crisis is threatening to turn into a full-scale depression, is important and timely. The current global economic crisis has placed the problem squarely on the forces promoting the neoliberal capitalist agenda across the world. And as the new forces that are emerging transform the global economy from private accumulation to collective responsibilities and action to correct the deteriorating situation, one can begin to see the seeds of alternative possibilities being planted to replace neoliberal capitalism with a new system. Going beyond the limits imposed by capitalism over the course of the twentieth century, one can anticipate the demands of the masses the world over to move us toward a new global social order that develops increasingly in the direction of socialism.

Introduction:
Globalization in the Twenty-first Century

Berch Berberoglu

Over one hundred years ago, in 1902, the eminent British scholar and liberal Member of Parliament, John A. Hobson, published a controversial book titled *Imperialism: A Study* (Hobson 1964 [1902]). A few years later, in 1916, relying in good part on Hobson's insightful observations on British imperialism, as well as Karl Marx's historical analysis of the development of capitalism through its various stages of concentration and centralization of capital and its accumulation on a world scale, Vladimir I. Lenin published a provocative exposé of global capitalism in the age of imperialism, titled *Imperialism: The Highest Stage of Capitalism*, to explain the dynamics of monopoly capitalism operating on a global scale in the early twentieth century (Lenin 1971 [1916]). Today, nearly a century later, we find ourselves in the midst of an intense debate on the relationship between capitalist imperialism of the early- to mid-twentieth century and neoliberal capitalist globalization of the late twentieth and early twenty-first centuries (Berberoglu 2005; 2009).

In their pioneering study of the evolution of neoliberal capitalist globalization of the past few decades, James Petras and Henry Veltmeyer put to rest this debate in a powerful critique of globalization studies that they published under the title *Globalization Unmasked: Imperialism in the 21st Century* (Petras and Veltmeyer 2001). It is within the context of Marxist critiques of neoliberal globalization, such as that provided by Petras and Veltmeyer, and political actions of popular social movements protesting against this phenomenon, that a new wave of discussion and debate on this topic has generated further study of the nature of globalization in the age of imperialism. This book attempts to further our understanding of this all-pervasive phenomenon by presenting a series of essays that address the many aspects of the globalization process and examine its dynamics and contradictions that have given rise to social movements that are now struggling against it.

Globalization is a complex phenomenon and social scientists have, over the course of the past few decades, defined it in various ways. My own studies of capitalist globalization, beginning with *Labor and Capital in the Age of Globalization* (2002) and subsequently in *Globalization of Capital and the Nation-State: Imperialism, Class Struggle, and the State in the Age of Global Capitalism* (2003), culminating in *Globalization and Change: The Transformation of Global Capitalism* (2005), have led me to define neoliberal globalization as "the highest stage of capitalist imperialism" (Berberoglu 2005, ix). The speed and intensity with which contemporary global capitalism has facilitated the accelerated expansion of capital on a world scale came to confirm my understanding of the fundamentals of modern capitalism encompassing the scope and depth of a process of exploitation and accumulation on a global scale that extended the capitalist control and domination of the world economy and the subjugation of the working class and all of humanity to the dictates of capital, the capitalist class, and the capitalist state on a worldwide basis throughout the course of the twentieth century and into the twenty-first.

What is globalization? What is its logic and mode of operation? What are its main characteristics? Who are its beneficiaries and who are the victims in this process that has been unfolding at great speed over the past several decades? Who, which groups and classes, are facilitating its expansion and who are the leading forces that are mobilizing against it? What are the dynamics of this process and its contradictions? And how will it continue to evolve and eventually become transformed? These are some of the key questions that many have been asking and debating over the past decade. And these and other related questions are what we propose to take up in this book.

Much of the recent *critical* literature on globalization that has developed over the past decade were inspired by the writings of Karl Marx, Rosa Luxemburg, and Vladimir I. Lenin, who were pioneers in the critique of the capitalist system as it developed in the nineteenth and early twentieth centuries. Marx, with his writings on British rule in India and Northern Ireland (1965 [1853]), Luxemburg, on the accumulation of capital from the national to the international level (1951 [1913]), and Lenin, with his analysis of the rise of monopoly capitalism to the global stage, where monopolies, trusts, and cartels came to rule over the world capitalist system (1971 [1916]), discussed in detail the rise of capitalism from the competitive market-based national economies to the monopoly-driven global economy dominated by industrial and subsequently finance capital, as capitalism became transformed into capitalist imperialism, that is, monopoly capital operating on a global scale, dominating the global economy, and dictating its terms against all its rivals.

Although, as Marx and Engels had earlier observed, "the executive of the modern [capitalist] State is but a committee for managing the common affairs of the whole bourgeoisie" (1972, 37), Lenin noted by the turn of the twentieth century that the rise of the monopolies and the rule of finance capital and its dictates over the capitalist class as a whole transformed the capitalist state to a state of the biggest and most powerful capitalist corporations, trusts, and cartels (i.e., the wealthiest segment of the capitalist class that owned and controlled them). Thus, the capitalist state, dominated by the monopolies, was forced to support and advance the latter's imperialist interests throughout the world. It is to this development—when competitive market-capitalism had given way to monopoly capitalism, that is, to capitalist imperialism—that even liberals such as John Hobson had come to object the policies and practices of imperialism as being detrimental to the national interest, and in Marxist terms detrimental to the interests of workers throughout the world—hence Marx's call for "workers of the world, unite!" and Lenin's provocative observation in *Imperialism: The Highest Stage of Capitalism* that "imperialism [i.e., monopoly capitalism operating on a global scale] is the eve of the social revolution of the proletariat"! (Lenin 1971, 175).

Not long after Marx and Lenin's path-breaking observations on the development of capitalism over the previous century, Marxist scholars such as Paul Baran and Paul M. Sweezy began to further develop a critique of capitalism in the twentieth century—most notably, Sweezy with his *The Theory of Capitalist Development* (1942) and Baran with his *The Political Economy of Growth* (1957), culminating in their joint study *Monopoly Capital* (1966). Extending this analysis to the global level, Harry Magdoff with his now-classic *The Age of Imperialism* (1969) and Andre Gunder Frank with his *Capitalism and Underdevelopment in Latin America* (1967) set the stage for Samir Amin's *Accumulation on a World Scale* (1974) and *Unequal Development: An Essay on the Social Formations of Peripheral Capitalism* (1976), which culminated in the emergence of what came to be known as "the dependency theory" of imperialism and underdevelopment that addressed the nature, contradictions, and consequences of imperialist expansion of the advanced capitalist economies beyond their boundaries, dominating and controlling the economies and societies of the less developed dependent periphery of the world capitalist system (Berberoglu 1992).

While these later left critics of early capitalist globalization viewed this as a process adversely affecting third world development, others within the Marxist tradition, such as Geoffrey Kay and Bill Warren, argued that this process is an outcome of the "normal" evolution of capitalist development, hence an inevitable consequence of capitalist expansion on a world scale (Kay 1975; Warren 1973, 1980). Thus, while Kay argued that "capital created underdevelopment not because it exploited the underdeveloped world, but because it did not exploit it enough" (1975: x), Warren went a step further by arguing that imperialism is the pioneer of capitalism—that is, through capitalist expansion across the globe, imperialism spreads capitalist relations of production throughout the world, hence the exploitation of labor on a global scale (Warren, 1980).

It is within the context of this divergence in left thought that we saw the emergence in the late 1970s and early 1980s of the "world systems" perspective (Wallerstein 1979) and the resurgence of classical Marxism on imperialism (Szymanski 1981). A synthesis of these two lines of thought by James Petras during this period, first in his *Critical Perspectives on Imperialism and Social Class in the Third World* (1978) and subsequently in his *Class, State and Power in the Third World* (1981), served as a basis for globalization studies that were yet to come in the 1990s and beyond. It is also during this critical phase of studies in international political economy that I began to address the problems of imperialism and globalization that we have come to discuss and debate today.

My initial study of the dynamics and contradictions of capitalist development in the twentieth century became crystallized in my 1987 book *The Internationalization of Capital: Imperialism and Capitalist Development on a World Scale*, which set the stage for my later studies of the globalization process, including *The Political Economy of Development* (1992) and *The Legacy of Empire: Economic Decline and Class Polarization in the United States* (1992). These and other works in the 1990s were the precursors of critical globalization studies of the past decade that came to dominate the social sciences, including the discipline of sociology.

I take great pleasure in compiling the present volume, parts of which originally emerged as a special issue of the journal *International Review of Modern Sociology* that I was invited to edit as guest editor to address the topic of globalization. In addition to several articles included here from that project, several more have been added to expand the project into book form for wider distribution worldwide. Hence, the book opens with a fresh new look at the dynamics of twenty-first-century globalization to set the stage for a critical understanding of both the neoliberal globalization project that is now in crisis and decline and the new forms of globalization that are in the making and will become prominent in the decades ahead.

Examining these historic and possible future developments in twenty-first-century globalization, Jan Nederveen Pieterse takes up these issues and provides ample empirical evidence on trends in the globalization process over the past two decades, with an analysis of the current global financial crisis and what may be in store for us in the immediate and near future. Following this survey of the contemporary predicament of global capitalism with projections for the future course of development of the globalization process itself, subsequent chapters in the book take up historical and case studies of the nature and dynamics of this phenomenon and its evolution right up to the twenty-first century.

Alan Spector in the next chapter contextualizes contemporary developments in capitalist globalization by taking up the challenge of tracing the origins and evolution of this process over the course of modern Western history, highlighting the evolution of globalization and imperialism through its various stages to the present. Spector argues that while changes have taken place in modern capitalism over the course of the twentieth century, capitalist business cycles and the globalization of capital today are fundamentally a product of the capital accumulation process on a worldwide basis and an extension of capitalist imperialism that serves the interests of the very same class forces in capitalist society.

Next, James Petras and Henry Veltmeyer dissect the neoliberal globalization project and examine its fault lines in Latin America. As the contradictions embedded in capitalist globalization generates the inevitable response, and radical rupture, against it—in this case in a most critical region of the world, Latin America (the "backyard" of U.S. imperialism)—a closer look at Latin America's integration into the globalization process reveals its complete failure that led to the demise of the neoliberal agenda that was ultimately rejected across the region.

As in other regions of the globe, Africa too has become affected by the impact of neoliberal globalization that is becoming increasingly prevalent across this region. Focusing on the marginalization of labor, women, and other exploited and oppressed segments of society by this process, Johnson W. Makoba next addresses the problems of neoliberal globalization as it affects Africa, especially when the transnationals are making great headway in their attempts to penetrate new territories, such as Africa.

Examining the contradictions and crises of global capitalism, the transnational corporations, and the imperial state, I take up an analysis of the relationship between imperialism and globalization in class terms and provide an account of the role of the imperial state in managing the affairs of the transnational capitalists, and doing so within the context of the parameters of the global capitalist system. Confined to the internal boundaries of global capitalism and its contradictions, global capital, I argue, is in the midst of a major crisis that it cannot get out from, and that this crisis—the deepest and most severe since the Great Depression—will affect the future of global capitalism and the globalization process for decades to come.

As societies outside the framework of the global capitalist system become integrated into it, as China has most recently, it becomes a process that is of interest in terms of its implications for future national and global developments. Alvin So, in his study of what he calls "the developmentalist state" in China, provides an interesting set of observations on the nature and contradictions of this process, highlighting the inner logic and possible consequences of this path of development in the era of globalization.

The impact of the broader process of neoliberal globalization, especially as it affects women in both the formal and informal economy, is taken up next by Lourdes Benería, as she focuses on the various ways in which women participating in wage-labor in export processing zones and informal employment deal with conditions they face in confronting the gendered nature of global social relations at the local level. How working women are able to deal with the forces that dominate their lives is a topic that is taken up in this important study that serves as a guide to similar such studies affecting women in various regions of the globe under conditions of neoliberal globalization.

Finally, Martin Orr, in his chapter on the contradictions and crisis of neoliberal capitalist globalization, argues that the failure of both the neoliberal and the neoconservative agenda signals the end of empire, and this in good part is the result of the resurgence of the anticapitalist-globalization movement that is succeeding in its worldwide struggle against transnational corporate globalization and capitalist imperialism.

The concluding chapter sums up the experience of neoliberal capitalist globalization across the globe, highlighting the nature and contradictions of this process to understand the inherent structural defects of the globalization process that explains its decline and fall, accompanying the decline and fall of Empire in the early decades of the twenty-first century. Contrasting emergent forms of alternative social organization to the failing neoliberal order at the national and transnational level, the concluding chapter explores possible paths out of the current crisis of global capitalism through mass mobilization and struggle.

Together, the ten chapters included in this book provide an important framework for further discussion and debate on the nature, dynamics, and contradictions of globalization and imperialism, historically and today, especially as we venture into a new and uncharted future in the unfolding twenty-first century.

References

Amin, Samir. 1974. *Accumulation on a World Scale*. New York: Monthly Review Press.

———. 1976. *Unequal Development: An Essay on the Social Formations of Peripheral Capitalism*. New York: Monthly Review Press.

Baran, Paul. 1957. *The Political Economy of Growth*. New York: Monthly Review Press.

Baran, Paul and Paul M. Sweezy. 1966. *Monopoly Capital*. New York: Monthly Review Press.

Berberoglu, Berch. 1987. *The Internationalization of Capital: Imperialism and Capitalist Development on a World Scale*. New York: Praeger Publishers.

———. 1992. *The Political Economy of Development: Development Theory and the Prospects for Change in the Third World*. Albany: State University of New York Press.

———. 2002. *Labor and Capital in the Age of Globalization: The Labor Process and the Changing Nature of Work in the Global Economy*. Boulder, CO: Rowman and Littlefield.

———. 2003. *Globalization of Capital and the Nation-State: Imperialism, Class Struggle, and the State in the Age of Global Capitalism*. Boulder, CO: Rowman and Littlefield.

———. 2005. *Globalization and Change: The Transformation of Global Capitalism*. Lanham, MD: Lexington Books.

Berberoglu, Berch. 2009. *Class and Class Conflict in the Age of Globalization.* Lanham, MD: Lexington Books.

Frank, Andre Gunder. 1967. *Capitalism and Underdevelopment in Latin America.* New York: Monthly Review Press.

Hobson, John A. 1964. [1902]. *Imperialism: A Study.* Ann Arbor: University of Michigan Press.

Kay, Geoffrey. 1975. *Development and Underdevelopment: A Marxist Analysis.* New York: St. Martin's Press.

Lenin, V. I. 1971. [1916]. "Imperialism: The Highest Stage of Capitalism," in V. I. Lenin, *Selected Works in One Volume.* New York: International Publishers.

Luxemburg, Rosa. 1951. [1913]. *Accumulation of Capital.* Reprint. New Haven: Yale University Press.

Magdoff, Harry. 1969. *The Age of Imperialism.* New York: Monthly Review Press.

Marx, Karl. 1965. [1853]. *On Colonialism.* New York: International Publishers.

Marx, Karl and Frederick Engels. 1972. "Manifesto of the Communist Party," in Karl Marx and Frederick Engels, *Selected Works.* New York: International Publishers.

Sweezy, Paul M. 1942. *The Theory of Capitalist Development.* New York: Monthly Review Press.

Szymanski, Albert J. 1981. *The Logic of Imperialism.* New York: Praeger.

Wallerstein, Immanuel. 1979. *The Capitalist World Economy.* Cambridge: Cambridge University Press.

Warren, Bill. 1973. "Imperialism and Capitalist Industrialization." *New Left Review* 81, September/October.

———. 1980. *Imperialism: Pioneer of Capitalism.* London: Verso.

2

Dynamics of Twenty-first-Century Globalization: New Trends in Global Political Economy

*Jan Nederveen Pieterse**

The twenty-first-century momentum of globalization is markedly different from twentieth-century globalization and involves a new geography of trade, weaker hegemony, and growing multipolarity. This presents major questions. Is the rise of East Asia, China, and India just another episode in the rise and decline of nations, another reshuffling of capitalism, a relocation of accumulation centers without affecting the logics of accumulation? Does it advance, sustain, or halt neoliberalism? The rise of Asia is codependent with neoliberal globalization and yet unfolds outside the neoliberal mold. What is the relationship between zones of accumulation and modes of regulation? What are the ramifications for global inequality? The first part of this chapter discusses trends in trade, finance, international institutions, hegemony and inequality, and social struggle.

The second part discusses what the new trends mean for the emerging twenty-first-century globalization and reflects on ramifications of the ongoing global economic crisis.

*I have presented various versions of this chapter at several institutions in fall 2006 (Korean Sociological Association conference, Seoul; Chulalongkorn University, Bangkok; Yunnan University, Kunming; the Chinese Academy of Social Sciences, Beijing; Globalism Institute, Royal Melbourne Institute of Technology; Jawaharlal Nehru University, New Delhi) and 2007 (Global Studies Association, University of California, Irvine) and am indebted to participants' feedback and the advice of many colleagues.

The financial crisis that erupted in 2008 and led to global recession in 2009 is part of the twenty-first-century transition and confirms several trends discussed below: the crisis of neoliberalism and American capitalism, weakening American hegemony, finance as a central arena of international competition, and the rise of emerging societies, in particular China, exemplified in the shift from the G7 to G20. The closing section of this chapter reflects on the significance of the crisis.

Twenty-first-Century Globalization

With 4 percent of the world population, the United States absorbs 25 percent of world energy supplies, 40 percent of world consumption, and uses up 50 percent of world military spending and 50 percent of world health care spending (at $1.3 trillion a year). U.S. borrowing of $700 billion per year or $2.6 billion per day absorbs 70 to 80 percent of net world savings. Meanwhile the U.S. share of world manufacturing output has steadily declined, and the share of manufacturing in U.S. GDP, at 12.7 percent, is now smaller than that of the health care sector at 14 percent and financial services at 20 percent. This shrinking of the physical economy in the United States makes it unlikely that the massive American external debt can ever be repaid (Prestowitz 2005).

According to IMF estimates, China and India are expected to overtake the GDP of the world's leading economies in the coming decades. China is expected to pass the GDP of Japan in 2016 and of the United States by 2025. In 2005, China surpassed the United States as Japan's biggest trading partner, surpassed Canada as the biggest trading partner of the United States, and surpassed the United States as the world's top choice of foreign direct investment. If current trends continue, China will become the biggest trading partner of practically every nation. By 2025 the combined GDP of the BRIC—Brazil, Russia, India, and China—would grow to one-half the combined GDP of the G-6 countries (the United States, Japan, Germany, France, Italy, Britain). According to a Goldman Sachs report, the combined BRIC economies will surpass that of the G-6 group, and "China, India, Brazil and Russia will be the first-, third-, fifth-, and sixth-biggest economies by 2050, with the United States and Japan in second and fourth place, respectively." BRIC spending growth measured in dollars could surpass the G-6 countries' levels as early as 2009 (Whelan 2004).

Both these data sets are uncontroversial, almost commonplace, yet combining them raises major questions. How do we get from here to there and what does this mean for the course and shape of globalization in the twenty-first century?

The United States, Europe, and Japan rode the previous wave of globalization, notably during 1980–2000, but their lead in manufacturing, trade, finance, and international politics is gradually slipping. The United States set the rules in economics through the Washington Consensus, in trade, through the WTO, in finance, through the dollar standard and the IMF, and in security, through its hegemony and formidable military. Each of these dimensions is now out of whack. The old winners are still winning, but the terms on which they are winning cedes more and more to emerging forces. In production and services, education and demography, the advantages are no longer squarely with the old winners. In several respects, in the maelstrom of globalization, the old winners have become conservative forces.

The twenty-first-century momentum of globalization is markedly different from twentieth-century globalization. Slowly, like a giant oil tanker, the axis of globalization is turning from North-South to East-South relations. This presents major questions. Is the rise of Asia and the newly industrialized economies (NIEs) just another episode in the rise and decline of nations, another reshuffling of capitalism, a relocation of accumulation centers without affecting the logics of accumulation? Does it advance, sustain, or halt neoliberalism? Is it just another shift in national economic fortunes, or is it an alternative political economy with different institutions, class relations, energy use, and transnational politics? What is the relationship between zones of accumulation and modes of regulation and what are the ramifications of these developments for global inequality?

Examining this poses methodological problems. It is risky to extrapolate trends. The units of analysis are not what they used to be or seem to be. Statistics measure countries, but economies are transborder phenomena. The story, of course, is not merely one of change, but also of continuity, and in some respects, seeming continuity.

Euro parliamentarian Glyn Ford notes, "The EU has more votes in the International Monetary Fund than the US, but has not yet used them to challenge the current neoliberal orthodoxy... With support from Latin America, in the World Trade Organization, at UN conferences in Tokyo as well as from the Santiago-plus-five and Durban-plus-five groupings, an alternative world could emerge" (Ford 2005).

It could, but so far it hasn't. There is a certain stickiness and stodginess to social change. Power plays continue as long as they can. Policies continue in the old style until a policy paradigm change is inevitable. There is a sleepwalking choreography to social existence, never quite in sync with actual trends; or rather, trends are only trends when they enter discourse. (In a similar way, what we teach in universities is often years behind what we know or what we're thinking about, because there is yet no convenient structure or heading under which to place and communicate it.) Changes manifest after certain time lags—an institutional lag, discursive lag, policy lag—yet changes get underway even if the language to signal them isn't quite there yet. Some changes we can name, some we can surmise, and some escape detection and will catch up with us. So at times it feels much like business as usual. Thus we should identify structural trends and discursive changes as well as tipping points that would tilt the pattern and the paradigm.

According to Kemal Dervis, director of the UN Development Program, globalization in the past was a profoundly "unequalizing process," yet "today, the process is rapidly turning on its head. The South is growing faster than the North. Southern companies are more competitive than their northern counterparts....Leading the charge is a new generation of southern multinationals, from China, Korea, India, Latin America and even the odd one from Africa, aggressively seeking investments in both the Northern and Southern Hemispheres, competing head-to-head with their northern counterparts to win market share and buy undervalued assets" (quoted in Peel 2005). This optimistic assessment counts economic changes—which this chapter also highlights—but it doesn't address social questions.

About cutting-edge globalization, there are two big stories to tell. One is the rise of Asia and the accompanying growth of East-South trade, energy, financial and political relations. Part of this story is being covered in general media, often with brio (Marber 1998; Agtmael 2007). In the words of Paul Kennedy, "we can no more stop the rise of Asia than we can stop the winter snows and the summer heat" (2001, 78). The other story, which receives mention only in patchy ways, is that emerging societies face major social crises in agriculture and urban poverty.

In the next section of this chapter I will discuss the main trends in twenty-first-century globalization by comparing trends during the periods of 1980–2000 and 2000–present under the headings of trade, finance, international institutions, hegemony, and inequality and social struggle. I discuss each of these at some length. In a subsequent section, I seek to understand what the new trends mean for the emerging twenty-first-century globalization.

Trade

Through the postwar period, North-South trade relations were dominant. In recent years, East-South trade has been growing, driven by the rise of Asian economies and the accompanying commodities boom (particularly since 2003) and high oil prices (since 2004). According to the UN Conference on Trade and Development, a "new geography of trade" is taking shape: "The new axis stretches from the manufacturing might and emerging middle classes of China, and from the software powerhouse of India in the South, to the mineral riches of South Africa, a beachhead to the rest of the African continent, and across the Indian and Pacific oceans to South America which is oil-rich and mineral- and agriculture-laden" (Whelan 2004).

Brazil opened new trade links with the Middle East and Asia. Chile and Peru are negotiating trade agreements with China (Weitzman 2005). "The Middle East has started looking to Asia for trade and expertise"; trade has expanded threefold in the past years, and the fastest growing markets for oil are in China and India (Vatikiotis 2005). Growing Sino-Indian trade combines countries with 1.3 and 1.2 billion people each (Dawar 2005).

During 1980–2000, American-led trade pacts such as NAFTA, APEC, and the WTO played a dominant role. In the 2000s these pacts are in impasse or passé. Dissatisfaction with NAFTA is commonplace, including in the United States. In Latin America, Mercosur, enlarged with Venezuela and with Cuba as associate member, undercuts the Free Trade Association of the Americas (FTAA). The association of Southeast Asian nations, ASEAN, in combination with Japan, South Korea, and China (ASEAN+3) sidelines APEC, which is increasingly on the backburner, and this reduces Asian dependence on the American market. Michael Lind (2005) notes, "This group has the potential to be the world's largest trade bloc, dwarfing the European Union and North American Free Trade Agreement".

During 1980–2000, the overall trend was toward regional and global trade pacts. The G22 walkout in Cancún in November 2003 upped the ante in subsequent negotiations. Advanced countries that previously pushed trade liberalization now resist liberalizing trade and retreat to "economic patriotism." The United States has been zigzagging in relation to the WTO (with steel tariffs and agriculture and cotton subsidies). Given the WTO gridlock in the Doha development round and blocked regional trade talks (the Cancún walkout was followed by the failure of the FTAA talks in Miami), the United States increasingly opts for free-trade agreements, which further erodes the WTO (Nederveen Pieterse 2004b).

There has been a marked shift toward bilateral free trade agreements (FTAs) in North-South trade. American terms in free-trade agreements typically include cooperation in the war on terror, exempting American forces from the International Criminal Court, accepting genetically modified food, and preferential terms for American multinationals and financial institutions. FTAs have been concluded with Chile, Colombia, Central America, Jordan, Morocco, Oman, and Singapore and are under negotiation with South Korea, Thailand, Australia, Peru, and Panama. In South-South trade, however, the trend is toward regional and interregional combinations, such as Mercosur and ASEAN. China has established a free-trade zone with ASEAN. In the future, India may join ASEAN+3. Since 2003, there have been talks to establish a free trade zone including India, Brazil, and South Africa (IBSA).

So the old "core-periphery" relations no longer hold. The South no longer looks to just North but also looks sideways. In development policies, East Asian and Southeast Asian models have long overtaken Western development examples. South-South cooperation, heralded as an alternative to dependence on the West ever since the Bandung meeting of the Nonaligned Movement in 1955, is now taking shape. "Already 43 percent of the South's global trade is accounted for by intra-South trade" (Gosh 2006, 7).

The downside is that much of this growth is sparked by a commodities boom that will not last. Note, for instance, the rollercoaster experience of the Zambian copper belt (Ferguson 1999), which now experiences another upturn, spurred by Chinese investments, which is as precarious as the previous round. Only countries that convert commodity surpluses into productive investments and "intellectual capital" will outlast the current commodities cycle. The commodities boom ended in 2008 and will not easily rebound.

Finance

During 1980–2000, finance capital played a key role in restructuring global capitalism. The financialization of economies (or the growing preponderance of financial instruments) and the hegemony of capital reflect the maturation of advanced economies, the role finance as a key force in globalization, financialization as the final stage of

American hegemony, and financial innovations, such as hedge funds and derivatives. The return to hegemony of finance capital is one of the defining features of neoliberal globalization (Duménil and Lévy 2001).

The role of speculative capital led to diagnoses such as casino capitalism and Las Vegas capitalism. International finance capital has been crisis prone, and such financial crises have hit Mexico, Asia, Russia, Latin America, and Argentina. Attempts to reform the architecture of international finance have come to little more than one-sided pleas for transparency. The trend since 2000 is that NIEs hold vast foreign reserves to safeguard against financial turbulence; "the South holds more than $2 trillion as foreign exchange reserves" (Gosh 2006, 7). As many historians note, the final stage of hegemony is financialization. Accordingly, emerging economies view competition in financial markets as the next strategic arena—beyond competition in manufacturing, resources, and services.

During 1980–2000 the IMF was considered as the hard taskmaster of developing economies; now, year after year, the IMF warns that U.S. deficits threaten global economic stability (Becker and Andrews 2004, Guha 2007), and by 2008 this came true.

Through the postwar period the U.S. dollar led as the world reserve currency, but since 2001 there has been a gradual shift from the dollar to other currencies. After the decoupling of the dollar from gold in 1971, OPEC in 1975 agreed to sell oil for dollars and established a de facto oil-dollar standard. Now Venezuela, Iran, and Russia price their oil in other currencies. In 2001–2005 the dollar declined by 28 percent against the euro, and by a further 12 percent in 2006. In 2002 the leading central banks held, on average, 73 percent of world reserves in dollars, by 2005 this was down to 66 percent (Johnson 2005), and the current trend is its further lowering toward 60 percent. China and Japan, with 70 to 80 percent of their foreign reserves in U.S. dollars, reflecting their close ties to the American market, deviate markedly from the world average. The recent trend shows China diversifying its foreign reserves and lowering its dollar reserves toward 65 percent (McGregor 2006). For obvious reasons this diversification must be gradual.

In the wake of the 1997 Asian crisis, the IMF vetoed Japan's initiative for an Asian monetary fund. Since then, Thailand's Chiang Mai Initiative established an Asian Bond Fund. Venezuela, backed by petrol funds, has withdrawn from the IMF and World Bank and has established an alternative Bank of the South. Japan, China, and South Korea, *if* they are able to settle their differences, might develop a yen-yuan-won Asian reserve, or an "Asian dollar." In 2009, Russia and China made proposals for an alternative to the dollar as world currency.

Western financial markets have been dominant since the seventeenth century. In the 2000s, financial sources outside the West began to play an increasingly important role, reflecting the rise of Asia, the global commodities boom, and high oil prices. The accumulation of petro money during 2005–2007 is three times the annual Asian surpluses from exports (Magnus 2006). A new east-east financial network is emerging. China's initial public offerings are increasingly no longer routed via New York and London, but via Saudi Arabia (Timmons 2006) and the Dubai Borse. Wall Street is losing its primacy to London as the center of world finance, with Shanghai and Hong Kong as runners up (Tucker 2007).

East Asian countries are active investors in Latin America and Africa. Thirty seven percent of FDI in developing countries now comes from other developing countries. China emerges as a new lender to the developing countries, at lower rates and without the conditions of the Washington institutions (Parker and Beattie 2006). China's foreign aid competes with Western donors, and Venezuela plays a similar role in Latin America.

Hedge funds have become more active international players than investment banks. In 2006 there were 10,000 hedge funds with $1.5 trillion in assets, the daily global turnover in derivatives was $6 trillion and the credit derivative market was worth $26 trillion. Financialization has increased the risk of financial instability (Glyn 2006), and new financial instruments such as derivatives are increasingly opaque and out of control. This underlies the financial instability that increasingly affects institutions in the West, such as the collapse of LTCM (Long Term Capital Management) in 1998, the Enron episode along with WorldCom, HealthSouth, and other corporations in 2001, Parmalat in 2003, Amaranth in 2006, and the crisis of American subprime mortgage lenders such as New Century in 2007 and 2008. This produced ripple effects throughout the global financial system. The deeper problem is that many American economic successes have been enabled by the Greenspan regime of easy money. An analyst comments, "This confusion of talent with temporary favorable conditions has combined to make clients willing to pay disproportionate fees" (Grantham 2007).

In the Davos meetings of the World Economic Forum the American economy and the unstable dollar have been a major cause of concern.

U.S. treasury debt at $7.6 trillion and net external debt at $4 trillion add up to an annual borrowing need of $1 trillion, or 10 percent of GDP (Buckler 2005), and interest payments of $300 billion a year and rising. The United States is deeply in the red to Asian central banks and relies on inflows of Asian capital and recycled oil dollars, and "what flows in could just as easily flow out" (Williams 2004). The dollar is now upheld more because of fear of turbulence than owing to its confidence and appeal. The Obama administration's deficit spending adds to the burdens.

For all these changes, the net financial drain from the global South is still ongoing. Poorer nations sustain American overconsumption and the overvalued dollar. The world economy resembles a giant Ponzi scheme with massive debt that is sustained by dollar surpluses and vendor financing in China, Japan, and East Asia. The tipping points are that financialization backfires when it turns out that financial successes (leveraged buyouts, mergers and acquisitions, and the rise in stock ratings) were based on easy credit, and secondly, when finance follows the "new money" in surplus countries.

Institutions

The 1990s institutional architecture of globalization was built around the convergence of the IMF, World Bank, and WTO, and is increasingly fragile. Since its handling of the Asian crisis in 1997–1998 and Argentina's crisis in 2001, the IMF has earned the nickname "the master of disaster." Argentina, Brazil, Venezuela, South Africa, Russia, and other countries have repaid their debt to the IMF early, so the IMF has less financial leverage, which is also because of the new flows of petro money. IMF lending went down from $70 billion in 2003 to $20 billion in 2006. The IMF has adopted marginal reforms (it now accepts capital controls and has increased the vote quota of several emerging economies) but faces financial constraints.

The World Bank has lost standing as well. In the 1990s the Bank shifted gear from neoliberalism to social liberalism and structural adjustment "with a human face," and an emphasis on poverty reduction and social risk mitigation. But the poverty reduction targets of the Bank and the Millennium Development Goals are, as usual, not being met. Paul Wolfowitz's attempts as World Bank president to merge neoliberalism and neoconservatism were counterproductive, with a divisive anticorruption campaign and focus on Iraq.

The infrastructure of power has changed as well. The "Wall Street-Treasury-IMF complex" of the nineties weakened because the

Treasury played a minor role in the G. W. Bush administration, until Henry Paulson's appointment in 2006, which brought Wall Street back in the cockpit. Since 2008, Wall Street has caved in, and since 2009 the expansion of IMF funds increasingly depends on emerging economies' contributions; in other words, the financial power structure has changed.

The 1990s architecture of globalization has become fragile for several reasons. The disciplinary regime of the Washington consensus has slipped away. Structural adjustment has shown a consistently high failure rate, with casualties in sub-Saharan Africa, most of Latin America, and the 1997 Asian crisis and the way it was handled by the IMF. Research indicates a correlation between IMF and World Bank involvement and negative economic performance, arguably for political reasons: since IMF involvement signals economic troubles, it attracts further troubles (McKenna 2005). Zigzag behavior by the hegemon—flaunting WTO rules, an utter lack of fiscal discipline, and building massive deficits—has further weakened the international institutions. Following the spate of financial crises in the nineties, crisis mismanagement, and growing American deficits, the macroeconomic dogmas of the Washington consensus has crumbled. Meanwhile, increasing pressure from the global South is backed by greater economic weight and bargaining power.

Hegemony

In general terms, the main possibilities in relation to hegemony are continued American hegemony, hegemonic rivalry, hegemonic transition, and multipolarity. The previous episode of hegemonic decline at the turn of the nineteenth century took the form of wars of hegemonic rivalry that culminated in the transition to the United States as the new hegemon. But the current transition appears to be structurally different from previous episodes. Economic and technological interdependence and cultural interplay are now far greater than at the fin de siècle. What is emerging is not simply a decline of (American) hegemony and rise of (Asian) hegemony but a more complex multipolar field.

During the 1990s, American hegemony was solvent, showed high growth, and seemed to be dynamic in the throttle of the new economic boom. The United States followed a mixed unipolar/multipolar approach with cooperative security (as in the Gulf War) and "humanitarian intervention" (as in Bosnia, Kosovo, and Kurdistan) as Leitmotivs. Unilateralism with a multilateral face during the 1990s gave way to unilateralism with a unilateral face under the G. W. Bush administration, a high-risk and high-cost approach that flaunted its weaknesses (Nederveen Pieterse 2008). By opting for unilateral "preventive war," the G. W. Bush administration abandoned international law. After declaring an "axis of evil," the United States had few tools left. The United States is now caught up in its new wars. In going to war in Iraq, the United States overplayed its hand. In its first out-of-area operation in Afghanistan, NATO encountered fierce resistance. The United States has been forced to give up its access to a base in Uzbekistan.

During the cold war, Muslims were cultivated as allies and partners on many fronts. Thus, in the 1980s, Ronald Reagan lauded the Mujahedeen in the Afghan war as "the moral equivalent of our founding fathers." As the cold war waned, these allies were sidelined. Samuel Huntington's "clash of civilizations" article in 1993 signaled a major turn by shifting the target from ideology to culture and from communism to the Islamic world. (In fact, he warned against a Confucian-Islamic alliance and specifically military cooperation between China and Pakistan.) Thus, the erstwhile allies and partners were redefined as enemies, and yesterday's freedom fighters were reclassified as today's terrorists.

In response to this policy shift and the continuing Israeli and American politics of tension in the Middle East, a militant Muslim backlash took shape, of which the attacks of September 11, 2001 were a part. The cold war "green belt" and "arc of crisis" has become an "arc of extremism" with flashpoints from the Middle East to Central Asia. Satellite TV channels in the Arab world contribute to awareness among Muslims. Muslim organizations increasingly demonstrate high militancy and swift responses, for instance, to the Danish cartoons and statements by Pope Benedict. The Lebanon war in 2006 showed Israel's weakness and Hezbollah's strength as part of a regional realignment away from the American supported Sunni governments to Iran, Syria, and the Shiites. The United States siding with Israel's insular stance in the region contributes to its self-isolation (Mearsheimer and Walt 2005; Petras 2006).

New security axes and poles have emerged, notably the Shanghai Cooperation Organization (deemed a "counterweight to NATO") and the triangular cooperation of China, Russia, and Iran. Other emerging poles of influence are India, Brazil, Venezuela, and South Africa. The G77 makes its influence felt in international trade and diplomacy.

For instance, it blocked intervention in Darfur on the grounds of state sovereignty, involving an Islamic government in a strategic part of the world, in part as a response to American expansion in the Middle East and Africa. China has generally backed G77 positions in UN Security Council negotiations (Traub 2006), a position that is now gradually changing.

On the military frontiers of hegemony, although the United States spends 48 percent of world military expenditure (in 2005) and maintains a formidable "empire of bases," the wars in Iraq and Afghanistan demonstrate the limits of American military power. As a traditional maritime and air power, the United States has traditionally been unable to win ground wars (Reifer 2005). "Globalization from the barrel of a gun" is a costly proposition, also because of the growing hiatus between American military and economic power (Nederveen Pieterse 2008).

On the economic front, the United States is dependent on imports and "Brand America" is losing points. In business circles the G. W. Bush presidency was viewed as a massive failure of American brand management. The aura of American power is fading. Rising anti-Americanism affects the status of American products, and American pop culture is no longer the edge of cool. An advertising executive notes growing resentment of American-led globalization:

> We know that in Group of 8 countries, 18 percent of the population claim they are avoiding American brands, with the top brand being Marlboro in terms of avoidance. Barbie is another one. McDonald's is another. There is a cooling towards American culture generally across the globe. (Holstein 2005)

The main tipping points of American hegemony are domestic and external. *Domestic* tipping points are the inflated housing market and high levels of debt. Not only are U.S. levels of debt high, but manufacturing capacity is eroded, there are no reserves and the domestic savings rate turned negative for the first time in 2005, so an adjustment is inevitable. If interest rates remain low, it undermines the appeal of dollar assets for foreign investors. If interest rates rise, it increases the pressure on domestic debt and the highly leveraged financial and corporate system. The main *external* tipping points are fading dollar loyalty, financial markets following new money, the growing American legitimacy crisis, and the strategic debacles in Iraq and the Middle East.

There are generally three different responses to American hegemony. The fist is *continued support*—which is adopted for a variety of reasons, such as the appeal of the American market, the role of the dollar, the shelter of the American military umbrella, and lingering hope in the possibility of American self correction. The second option is *soft balancing*—which ranges from tacit noncooperation (such as most European countries staying out of the Iraq war and declining genetically modified food) to establishing alternative institutions without U.S. participation (such as the Kyoto Protocol and the International Criminal Court). And the third response is *hard balancing*—which only few countries can afford, either because they have been branded as enemies of the United States already and so have little to lose (Cuba, Venezuela, Iran, Sudan), or because their bargaining power allows them maneuvering room (as in the case of China and Russia and the SCO). An intriguing trend is that the number of countries that *combines* these different responses to American hegemony in different policy domains is increasing. Thus, China displays all three responses in different spheres—economic cooperation (WTO, trade), noncooperation in diplomacy (UN Security Council) and finance (valuation of renminbi), and overt resistance in Central Asia (Wolfe 2005) and support for Iran.

American unilateralism and preventive war are gradually giving way to multipolarity, if only because unilateralism is becoming too costly, militarily, politically, and economically. New clusters and alignments are gradually taking shape around trade, energy, and security. Sprawling and cross-zone global realignments point to growing multipolarity rather than hegemonic rivalry.

Inequality and Social Struggle

Let us review these trends in a wider time frame. Postwar capitalism from 1950 to the 1970s combined growth and equity. Although overall North-South inequality widened, economic growth went together with growing equality among and within countries. Neoliberal "free market" economies during 1980–2000 produced a sharp break in this trend: now economic growth came with sharply increasing inequality within and among countries. The main exceptions to the trend were the East Asian tiger economies.

The trend in the 2000s is that overall inequality between advanced economies and emerging economies is narrowing while inequality in emerging societies is increasing. Overall global inequality is staggering, with 1 percent of the world population owning 40 percent of the world's assets. The pattern of rising inequality in neoliberal economies (the United States, the UK, and New Zealand) continues and has begun to extend to Australia, Japan, and South Korea (Lim and Jang 2006). International migration has become a major flashpoint of global inequality and produces growing conflicts and dilemmas around migration and multiculturalism in many countries (Nederveen Pieterse 2007).

James Rosenau offers an optimistic assessment of global trends, according to which rising human development indices, urbanization, and growing social and communication densities are producing a general "skills revolution" (1999). However, the flipside of technological change and knowledge economies is that with rising skill levels come widening skills differentials and urban-rural disparities. The second general cause of growing inequality is unfettered market forces promoted by transnational corporations, international institutions, and the corporate media. Familiar short hands are shareholder capitalism (in contrast to stakeholder capitalism), Wal-Mart capitalism (low wages, low benefits, and temp workers), and Las Vegas capitalism (speculative capital). The third general cause of inequality is financialization, because its employment base is much narrower than in manufacturing, and income differentials are much steeper. A fourth cause of inequality in developing countries is fast-growth policies that reflect middle-class and urban bias and aggravate rich-poor and urban-rural gaps.

Practically all emerging economies face major rural and agricultural crises. In China this takes the form of pressure on land, deepening rural poverty, pollution, village-level corruption, and urban migration. In Brazil and the Philippines, land reform drags because the political coalition to confront landholding oligarchies is too weak. In South Africa, the apartheid legacy and the poor soil and weak agricultural base in the former Bantustans contribute to rural crisis.

Yet the *impact* of poor peoples' movements and social struggles in the 2000s has been greater than during 1980–2000, notably in China and Latin America. In China, where "a social protest erupts every five minutes," social crises are widely recognized and have led to the "harmonious society" policies adopted in 2005. In Latin America, poor peoples' movements have contributed to the election of left-wing governments in Venezuela, Bolivia, Ecuador, and Nicaragua and to policy adjustments in Argentina and Chile.

Whereas the "Shanghai model" of fast-growth policies that are geared to attract foreign investment has been abandoned in China, it is being pursued with fervor in India. A case in point is the "Shanghaing of Mumbai" (Mahadevia 2006) and the growing role of special economic zones. What is the relationship between the India of Thomas Friedman (*The world is flat*) and P. Sainath (*Everybody loves a good drought*), between celebrating growth and deepening poverty, between Gurgaon's Millennium City of Malls and abject poverty kilometers away, between dynamic "Cyberabad" and rising farmer suicides nearby in the same state of Andhra Pradesh? According to official figures, 100,248 farmers committed suicide between 1993 and 2003. Armed Maoist struggles have spread to 170 rural districts, affecting 16 states and 43 percent of the country's territory (Johnson 2006), and it is now the country's top security problem.

> For every swank mall that will spring up in a booming Indian city, a neglected village will explode in Naxalite rage; for every child who will take wings to study in a foreign university there will be 10 who fall off the map without even the raft of a basic alphabet to keep them afloat; for every new Italian eatery that will serve up fettuccine there will be a debt-ridden farmer hanging himself and his hopes by a rope. (Tejpal 2006)

India's economic growth benefits a top stratum of 4 percent in the urban areas, with little or negative spin off for 80 percent of the population in the countryside. The software sector rewards the well educated middle class. The IT sector has an upper-caste aura—brainy, requiring good education, English language—and extends upper-caste privileges to the knowledge economy, with low-cost services from the majority population in the informal sector (Krishna and Nederveen Pieterse 2008). Public awareness in India is split between middle-class hype and recognition of social problems, but there are no major policies in place to address the problems of rural majorities and the urban poor.

In addition to rural crisis, the emerging powers face profound *urban poverty* as part of the "planet of slums" (Davis 2005). The rural crisis feeds into the sprawling world of the favelas, bidonvilles, shanty towns, and shacks. Urban policies are at best ambivalent to the poor and often negligent. Thus, Bangkok's glitzy monorail mass transit system connects different shopping areas, but not the outlying suburbs. As India's rural poor are driven out of agriculture, they flock to the cities, while land appropriations and clampdowns on informal settlements, hawking, and unlicensed stores squeeze the urban poor out of the cities, creating a scissor operation that leaves the poor with nowhere to go.

Trends in Twenty-first-Century Globalization

Now let us review these trends. Is the cusp of the millennium, 1980–2000 and 2000–present, a significant enough period to monitor significant changes in globalization? Why in a short period of decades would there be significant trend breaks? My argument is essentially that two projects that defined the 1980–2000 period, American hegemony and neoliberalism—which are, of course, the culminating expressions of longer trends—are now over their peak. They are not gone from the stage but they gather no new adherents and face mounting problems (indebtedness, military overstretch, legitimacy crises, rising inequality), and new forces are rising. The new forces stand in an ambiguous relationship to neoliberalism and American hegemony.

In sum, the overall picture shows distinct new trends in trade, institutions, finance, and hegemony and to some extent in social inequality. Table 1 reviews the main trends in current globalization. The trend break with the old patterns is undeniable, yet it is too early to speak of a new pattern.

We can also reflect on these changes in a longer time frame. According to the thesis of Oriental globalization (Hobson 2004; Nederveen Pieterse 2006), early globalization was centered in the Middle East (500–1100 CE) and between 1100 and 1800 it was centered in China, India, and Southeast Asia. Now, as a Shanghai economist remarks, after "a few hundred bad years" China and India are back as the world's leading manufacturing center and information processing center, respectively (Prestowitz 2005).

Thus, in a historical sense, twenty-first-century globalization is reverting to normal if we consider that Asia has been at the center of the world economy through most of the long-term globalization process. In this light, two hundred years of Western hegemony have been a historical interlude.

Note, for instance, that it is not the first time that China is in the position of having accumulated the lion's share of the world's financial reserves. During "several periods of rapid growth in international commerce—from A.D. 600 to 750, from 1000 to 1300, and from 1500 to 1800—China tended to run very large trade surpluses." Between 1500 and 1800 China accumulated most of the world's silver

Table 1   Trends in Twenty-first-Century Globalization

Pattern 1990s	Pattern 2000s
Trade	
North-South trade dominates	Growing East-South trade
U.S.-led trade pacts dominate	FTAA, APEC, WTO: passé or in impasse
Trend to regional/global trade pacts	Shift to bilateral FTAs (in North-South trade)
Finance	
Finance capital leads, crisis prone	Emerging economies hold dollar surpluses
IMF and World Bank discipline developing economies	IMF warns United States its policies threaten economic stability
U.S. dollar leads	Decline of dollar as world reserve currency
United States is top destination of FDI	China top destination of FDI
IMF blocks Asian monetary fund	Thai Asian Bond Fund; Bank of the South
Western financial markets dominate	New financial flows outside the West
Investment banks	Hedge funds, new financial instruments
Institutions	
Convergence IMF-WB-WTO	IMF lending down ($70bn 2003, $20bn 2006)
Social liberalism, poverty reduction	World Bank lost standing
"Wall Street-Treasury-IMF complex"	Weak Treasury
Washington consensus	(Post)Washington no-consensus
Hegemony	
U.S. hegemony solvent and dynamic	United States in deficit and cornered in new wars
"Clash of civilizations"	Muslim backlash
U.S.-led security	New security axes and poles
Inequality	
Growth & increasing inequality (except East Asia)	Inequality between North and NIEs decreases while inequality in NIEs increases
Deepening rural and urban poverty	Deepening rural and urban poverty International migration as flashpoint of global inequality

and gold (Bradsher 2006; Frank 1998). So it is not the first time in history that China faces the "trillion dollar question" of holding the world's largest financial surplus.

Now, however, Asia resumes its normal role in a world that is imprinted and shaped by two hundred years of Western hegemony—in politics, military affairs, corporate networks, intellectual property rights and patents, institutions, styles and images. Asia makes its comeback in a world that, unlike in 1800, is deeply interconnected socially, politically, and culturally, a world that is undergoing rapid technological change, more rapid than in 1800.

The West followed Asia and transcended it by introducing new forms of production (industrialism, mass production, Fordism), and now Asia follows the West and transcends it. Japan pioneered flexible accumulation, and the East Asian development states and the "Beijing consensus" represent other modes of regulation, and the question is which of the modes of regulation Asia introduces will prove to be sustainable.

According to American conventional wisdom and authors such as Thomas Friedman (2005), China's economic rise follows Deng's four modernizations and the subsequent liberalization, and India's economic rise dates from its 1991 liberalization. These views are ideology rather than research-based, because research indicates different itineraries. Rodrik's work on the "Hindu rate of growth" argues that the foundations of India's economic resurgence were laid during the 1970s and 1980s (2004). Recent studies of China break the mold of Mao stigmatization and find that improvements in industrial production, rural modernization, literacy, and health care during Mao's time laid the groundwork for the post-1978 transformation (Gittings 2005; Guthrie 2006).

Liberalization and export orientation—the Washington consensus and World Bank formulae—contributed to the rise of Asia. American offshoring and outsourcing have spurred rapid growth (Wal-Mart's imports alone represent 15 percent of the U.S. trade deficit with China; Prestowitz 2005, 68). But this would not have been possible or produced sustainable growth without Asia's developmental states. Their development policies enable Asian societies and producers to upgrade technologically and to foster domestic, regional, and alternative markets. China's spending on high-tech research and development now ranks third after the United States and Japan.

Alternatives that were sidelined during the epoch of neoliberal hegemony have taken on new influence and legitimacy since the turn of the millennium. The Beijing consensus—"a model for global development that is attracting adherents at almost the same speed that the U.S. model is repelling them" (Ramo 2004), is an emerging alternative in Asia, and the Bolivarian alternative has been gaining ground in Latin America. Countries that are financially independent and have relative maneuvering room, such as China because of its size and Venezuela because of its oil wealth, are in a strong position to articulate alternatives to neoliberalism.

If we look at the world as a whole, the majority economic form is the mixed economy with the social market in the EU, bureaucratically coordinated market economies (Japan), and developmental states (with different leanings in Asia, Latin America, and Africa). On balance, mixed economies are doing better and several are more sustainable in terms of their growth paths and energy use. Social market and human development approaches are generally coming back on the agenda. Global emancipation hinges on rebalancing the state, market, and society and introducing social cohesion and sustainability into the growth equation. Let me add brief notes on the significance of the ongoing economic crisis for twenty-first-century globalization.

Crisis and Twenty-first-Century Globalization

The crisis confirms several broad trends: American capitalism loses its leading role; finance is a central arena of international competition—as the Asian crisis already made clear (cf. Bello et al. 2000); China gradually assumes a pivotal global role. In broad strokes: international trade is down, finance cools off, and Asia and emerging societies are rising relative to the West. The crisis, in a sense, reflects and corrects the global imbalance that has built up during past decades, in brief: consumption and deficits in the United States, production and surpluses in Asia. Inevitably, then, the balance tilts toward Asia and the countries with surpluses.

The crisis, in one reading, is an expression of global imbalance. According to Krishna Guha, "the current crisis is in the strictest sense a crisis of globalization, fostered and transmitted by the rapid and deep integration of very different economies. Fast-growing developing countries with underdeveloped financial systems were exporting savings to the developed world for packaging and reexport to them in the forms of financial products... the claim that this was sustainable assumed core financial centres—above all New York and London— could create the financial products efficiently and without blowing up. They could not..." (Guha 2009). In this account, by implication, the culprit is the "savings glut" in Asia that has overwhelmed innocent American financial institutions. This narrative overlooks that three decades of deregulation had magnified the vulnerability of these institutions, including the Clinton administration's repeal of the Glass-Steagall Act (eliminating the barrier between commercial and investment banks), while the Federal Reserve's policies of low interest rates relayed the financial inflows through an easy money regime, creating a credit bubble society throughout America with mounting debt. The subprime mortgages were the latest and most vulnerable extension of this financial sandcastle.

That neoliberalism is unsustainable has long been argued (e.g., Nederveen Pieterse 2000). The crisis illustrates the combined effect of deregulation and financialization, two features of the neoliberal era. The financial sector amplified by financialization, the Wall Street power houses, is now referred to as "death-wish finance" (Crook 2009). The third component of neoliberalism, the exploitation of right-less labor and the concentration of power at the top, which I call Dixie capitalism (Nederveen Pieterse 2004a), is yet to be confronted.

Attempts at crisis management now include the new forces—emerging societies, surplus countries, and oil exporters—simply because only they have the resources that could restore balance to the global economy. The G20 meeting in April 2009 in London illustrates the new momentum and gave "unspoken recognition of the remaking of the geopolitical landscape" (Stephens 2009). The meeting, which actually included twenty-nine delegations, was a summit of "the rise of the rest." A new power balance is taking shape. The IMF obtains new funds to address the credit squeeze for developing countries—from newcomers to the world power structure, which thereby take their seat at the head table. The creditors point out terms and begin to exact conditions, such as China cautioning the U.S. government to maintain fiscal probity with a view to the security of its vast dollar holdings, and sovereign wealth funds monitoring their investments closely after having suffered huge losses on their earlier investments in American investment banks. However, without structural reforms, bailouts and crisis management will not succeed. Additional funds will simply go where the previous ones went, into the gigantic sinkhole of financial wizardry, derivatives, credit swaps, CDOs, and other arcane financial instruments, which share the features of opaqueness and deception. Saskia Sassen (2009) notes the magnitudes of financial bubbles—financial assets in the United States reached 4.5 times of GDP in 2008 and in the European Union 3.5 times; globally the value of debt stood at $160 trillion, three times the global GDP. She proposes *de-financialization* or bringing the financial sector in line with the "real" economy, a proposal as sweeping as Walden Bello's call for deglobalization (2003).

What is needed is unscrambling finance into different circuits: utility finance (retail and corporate banking, insurance, underwriting, foreign exchange) and speculative finance (hedge funds, trade in derivatives, credit swaps, and other arcane financial instruments) (Plender 2009; Jackson 2009). Casino operations caused the failure of financial houses from Barings to Lehmann and insurance firms such as AIG. What is needed then, first, is to reerect the barriers between circuits and corridors of finance, such as the Glass-Steagall Act in the United States that split commercial and investment banking. Second, *within* financial institutions there should be barriers between functions and circuits so investment banks don't play at gambling tables. The common principle is to establish fire doors so gambling operations can't bring the house down (as happened with Barings bank). Third, regulatory institutions and credit rating agencies should be redesigned accordingly. Fourth, the overall financial sector should be cut in size and fees brought in line with utility standards. Schumpeter contrasted the continental European tradition of public service banking to Anglo-American practices. These reforms would bring Anglo-American banking in line with public service functions. It would entail reforms of the kind envisaged by the French and German governments whose insistence on financial regulation, rather than bailouts, was shushed at the G20 meeting. This suggests the old power structure prefers a long rope rather than a short stick.

References

Agtmael, Antoine van. 2007. *The Emerging Markets Century.* New York: The Free Press.

Becker, E. and E. L. Andrews. 2004. "IMF Says Rise in U.S. Debts is Threat to World Economy," *New York Times*, August 1.

Bello, Walden. 2003. *Deglobalization: Ideas for a New World Economy.* London: Zed.

Bello, W., N. Bullard, and K. Malhotra, eds. 2000. *Global Finance: New Thinking On Regulating Speculative Capital Markets.* London: Zed.

Bradsher, Keith. 2006. "From The Silk Road to the Superhighway, All Coin Leads to China," *New York Times*, February 26: WK4.

Buckler, William A. M. 2005. "Global Report," *Privateer*, 518, January: 1–12.

Crook, Clive. 2009. "Strike Faster on Death-Wish Finance." *Financial Times*, March 23: 7

Davis, Mike. 2005. *Planet of Slums.* London: Verso.

Dawar, Niraj. 2005. "Prepare Now for a Sino-Indian Trade Boom," *Financial Times*, October 31: 11.

Duménil, G. and D. Lévy. 2001. "Costs and Benefits of Neoliberalism: A Class Analysis," *Review of International Political Economy* 8, 4: 578–607.

Ferguson, James. 1999. *Expectations of Modernity*. Berkeley: University of California Press.

Ford, Glyn. 2005. "Forging an Alternative to US Hegemony," *Japan Times*, February 7.

Frank, A. G. 1998. *Re Orient: Global Economy in the Asian Age*. Berkeley: University of California Press.

Friedman, Thomas L. 2005. *The World is Flat*. New York: Farrar Straus and Giroux.

Gittings, John. 2005. *The Changing Face of China: From Mao to Market*. Oxford: Oxford University Press.

Glyn, Andrew. 2006. "Finance's Relentless Rise Threatens Economic Stability," *Financial Times*, April 27: 13.

Gosh, Parthya S. 2006. "Beyond the Rhetoric," *Frontline*, October 6: 7–9.

Guha, Krishna. 2007. "IMF Warns of Risk to Global Growth," *Financial Times*, August 22: 3.

———. 2009. "Imbalances Imply a Trouble Well Beyond Risky Banking," *Financial Times*, March 10: 9.

Guthrie, Doug. 2006. *China and Globalization: the Social, Economic and Political Transformation of Chinese Society*. New York: Routledge.

Hobson, John M. 2004. *The Eastern Origins of Western Civilization*. Cambridge: Cambridge University Press.

Holstein, William J. 2005. "Erasing the Image of the Ugly American," *New York Times*, October 23: B9.

Jackson, Tony. 2009. "No Return To Playing The Tables," *Financial Times*, April 6, 2009: 17.

Johnson, Jo. 2006. "Insurgency in India—How the Maoist Threat Reaches Beyond Nepal," *Financial Times*, April 26: 13.

Johnson, Steve. 2005. "Indian and Chinese Banks Pulling Out of Ailing U.S. Dollar," *Financial Times*, March 7.

Kennedy, Paul. 2001. "Maintaining American Power: From Injury to Recovery," in S. Talbott and N. Chanda, eds, *The Age of Terror: America and the World After September 11*. New York: Basic Books, 53–80.

Krishna, Anirudh and J. Nederveen Pieterse. 2008. "Hierarchical Integration: The Dollar Economy and the Rupee Economy in India," *Development and Change* 39, 2: 219–237.

Lim, Hyun-Chin and Jin-Ho Jang. 2006. "Neoliberalism in Post-Crisis South Korea," *Journal of Contemporary Asia* 36, 4: 442–463.

Lind, Michael. 2005. "How the U.S. Became the World's Dispensable Nation," *Financial Times*, January 26.

Magnus, G. 2006. "The New Reserves of Economic Power," *Financial Times*, August 22: 11.

Mahadevia, D. 2006. *Shanghaing Mumbai: Visions, Displacements and Politics of a Globalizing City*. Ahmedabad: Centre for Development Alternatives.

Marber, Peter. 1998. *From Third World to World Class: The Future of Emerging Markets in the Global Economy*. Reading, MA: Perseus.

McGregor, Richard. 2006. "Pressure Mounts on China Forex Management," *Financial Times*, November 28: 6.

McKenna, Barrie. 2005. "With Friends like the IMF and the World Bank, Who Needs Loans," *Globe and Mail*, August 16: B11.

Mearsheimer, John and Stephen Walt. 2006. "The Israel Lobby," *London Review of Books*, March 23, 3–12.

Nederveen Pieterse, J. 2000. "Shaping Globalization," in J. Nederveen Pieterse, ed. *Global Futures: Shaping Globalization*. London, Zed Books, 1–19.

———. 2004a. *Globalization or Empire?* New York: Routledge.

———. 2004b. "Towards Global Democratization: To WTO or not to WTO?" *Development and Change* 35, 5: 1057–1063.

———. 2006. "Oriental Globalization: Past and Present," in G. Delanty, ed. *Europe and Asia Beyond East and West: Towards a New Cosmopolitanism*. London: Routledge, 61–73.

———. 2007. *Ethnicities and Global Multiculture: Pants for an Octopus*. Lanham, MD: Rowman & Littlefield.

———. 2008. *Is There Hope for Uncle Sam? Beyond the American Bubble*. London: Zed.

Parker, G. and A. Beattie. 2006. "Chinese Lenders 'Undercutting' on Africa Loans," *Financial Times*, November 29: 3.

Peel, Quentin. 2005. "The South's Rise is Hindered at Home," *Financial Times*, November 17: 17.

Petras, James. 2006. *The Power of Israel in the United States*. Atlanta, GA: Clarity Press.

Plender, John. 2009. "The Financial Sector Is Far From Being A Busted Flush," *Financial Times*, March 18: 22.

Prestowitz, Clyde. 2005. *Three Billion New Capitalists: the Great Shift of Wealth and Power to the East*. New York: Basic Books.

Ramo, Joshua Cooper. 2004. *The Beijing Consensus*. London: Foreign Policy Centre.

Reifer, T. E. 2005. "Globalization, Democratization, and Global Elite Formation in Hegemonic Cycles: A Geopolitical Economy," in J. Friedman and C. Chase-Dunn, eds. *Hegemonic Declines: Past and Present*. Boulder, CO: Paradigm.

Rodrik, Dani. 2004. "Globalization and Growth—Looking in the Wrong Places," *Journal of Policy Modeling*, 26: 513–517.

Rosenau, J. N. 1999. "The Future of Politics," *Futures* 31, 9–10: 1005–1016.

Sainath, P. 1996. *Everybody Loves a Good Drought*. New Delhi: Penguin.

Sassen, Saskia. 2009. "Too Big To Save: The End of Financial Capitalism." *Open Democracy News Analysis*, April 2.

Stephens, Philip. 2009. "A Summit Success That Reflects a Different Global Landscape." *Financial Times*, April 3: 9.

Tejpal, Tarun J. 2006. "India's Future, Beyond Dogma," *Tehelka, The People's Paper*, 25 November: 3.

Timmons, H. 2006. "Asia Finding Rich Partners in Middle East," *New York Times*, December 1: C1–5.

Traub, J. 2006. "The World According to China," *New York Times Magazine*, September 3: 24–29.

Tucker, S. 2007. "Asia Seeks its Centre," *Financial Times*, July 6: 7.

Vatikiotis, Michael. 2005. "Why the Middle East is Turning to Asia," *International Herald Tribune*, June 24.

Weitzman, Hal. 2005. "Peru Takes Faltering Steps in Bid to Win China prize," *Financial Times*, May 30.

Whelan, Caroline. 2004. "Developing Countries' Economic Clout Grows," *International Herald Tribune*, July 10–11: 15.

Williams, Ian. 2004. "The Real National Security Threat: the Bush Economy," *AlterNet*, January 13.

Wolfe, Adam. 2005. "The 'Great Game' Heats Up in Central Asia," *Power and Interest News Report*, August 3.

3

Neoliberal Globalization and Capitalist Crises in the Age of Imperialism

Alan J. Spector

During the past two decades, the concept of "globalization" has been used in reference to global social, economic, and political processes as if it were some profoundly new development in world history. In this chapter, I argue that while the scale and breadth of this recent wave of globalization is certainly unparalleled, the underlying polit-ical-economic processes that are its mode of operation have been in place for over a century, and have given rise to previous waves of glo-balization during the course of the twentieth century. Pre-nineteenth-century global movement of capital, technology, and people were driven by precapitalist colonial expansion. Under capitalism, espe-cially during the stage of industrial-finance capital over the past cen-tury, maintenance of rates of profit, rather than simply securing land, material resources, or slaves, has been the driving force of global eco-nomic expansion. Thus, I argue that understanding economic and sociopolitical processes within the framework of modern capitalist imperialism—including, especially, the need for cheap labor, raw materials, and new markets abroad, and political-military policies that protect these interests around the world—provides us a useful framework for understanding the nature, dynamics, and contradic-tions of this most recent wave of globalization and the various forms of resistance against it.

Before 1990, one had hardly heard of the word "globalization." The word, or variants of it, first appeared in the *Oxford English Dictionary* in 1961. In the most general sense, globalization refers to increasing worldwide integration of the economy, culture, technology, communication and information-sharing, and the movement of people. It also somewhat refers to political integration, although that process is more problematic. While it is trendy to use the term as if it is a new process that appeared suddenly, aspects of this process have been developing for thousands of years. This most recent trend is often called "neoliberal globalization," but in order to better understand its dynamics, it is important to examine earlier trends toward globalization and to delineate what makes it different from these earlier trends (Harvey 2005). Thus, one can analyze and categorize different types of globalization throughout history as the result of various political-economic systems, the most recent of which is based on the logic of capitalist imperialism.

Earlier Trends toward Globalization

In ancient times, there certainly were periods of broader integration of the economy, culture, technology (especially in agriculture), and the movement of people across the globe. Generally, it was the result of conquests that were driven by the search for more wealth and more people to subjugate and exploit. The term "globalization" is, of course, too much of a stretch to apply to these processes, but it is important to understand that the movement of peoples and the spread of technology are not new developments. Most social scientists now understand the term "Dark Ages" to be a very simplistic description of European, and certainly world, history for the time period between what is considered "the ancient world" and European exploration and conquest of the Western Hemisphere. During that time, there was extensive travel, exploration, and trade extending from the Pacific Coast of China all the way to Western Europe and North Africa. The recent best-selling book *Guns, Germs, and Steel*, has done much to popularize this knowledge, which had been taken for granted by historians and anthropologists for many years (Diamond 2005). Again, during that era, there were many waves of migration, many periods in which transcontinental and intercontinental movements of people and exchange of knowledge accelerated between Asia, Europe, and, to some extent, Africa. War has always been a particularly potent accelerant of these processes.

It could be argued that, since 1492, there has been a continuous process of increased "globalization," with periods of acceleration and deceleration, to be sure. Understanding these developments, however, requires more than just measuring the quantitative aspects of these processes. It is especially important to understand their qualitative nature—that is, what economic, political, and sociocultural forces shaped the kinds of globalization. For this reason, we may discuss three similar, but not identical, globalizations, understood in terms of these forces.

The first of these three was obviously sparked by the needs of various classes in Europe to exploit the wealth of the Western Hemisphere. While the word "imperialism" has been loosely used to describe every economic-military expansion, whether from Ancient Rome or sixteenth century Spain, late nineteenth-century Britain or the U.S. invasion and occupation of Afghanistan and Iraq in the early twenty-first century, it should be clear that those "imperialisms" might be superficially similar, but they are quite different in essence, because of the different underlying political-economic forces that generated and promoted these actions.

The accelerated migration of people and exchange of technology and culture that took place in the sixteenth, seventeenth, and early eighteenth centuries were not solely between Europe and the Americas. The enslavement and transport of millions of Africans to the Western Hemisphere should certainly be considered an event of world-changing significance. Trade (and war) between various European powers and various African groups as well as the movement of agricultural products by the Europeans between Latin America, Africa, and Asia all contributed to the "shrinking of the world." As Galeano and others have pointed out, the global interconnections were complex and profound during that period (Galeano 1997). For example, it was not mainly Spain that gained great wealth and power from the conquest of Latin America and the extraction of wealth from mining and agricultural labor. In fact, the Spanish nobility spent much of its wealth from Latin America importing products from Britain, which in turn invested a major part of its profits from that arrangement; South American mining and agriculture were significant factors in funding Britain's industrialization and consolidation of the British Empire. Galeano even argues that the research that led to the development of the steam engine by Scotland's James Watt was funded in large part by the exploitation of labor in the Americas by Spain. China, too, was a beneficiary of this situation, as the Spanish rulers had a passion for Chinese textiles and other products.

The "globalization" of the world economy is not something that started in the 1990s. The driving force for much of this was the development and growth of capitalism, which is much more dynamic, and often unstable, than earlier political-economic systems. Capitalism as the dominant system did not consolidate its power until the nineteenth century, primarily in Europe and North America, and by extension other areas of the world. But capitalist *processes*, the rise of the capitalist class, and the major trading companies and banks all grew significantly during the period before capitalism was consolidated. The constant need to accumulate the capital necessary for this class to become dominant and the need for profits to satisfy investors created a dynamic that was different from earlier periods in human history when wealth was accumulated largely to be consumed.

Imperialism and Modern Capitalism

The period from the early- to mid-1800s until the latter part of the twentieth century saw the development of a kind of globalization/ imperialism that shared important characteristics with the previous two or three centuries of capitalist wealth accumulation, but which was different in important ways. It is not necessary, nor would it be accurate, to postulate precise start and end dates for these periods. What is distinctive about this period is that capitalism, as a system, was now consolidated as the dominant political-economic system in much of the world, and in much of those parts of the world where precapitalist relations seemed to predominate on the local level, it was generally the case that the overall economies of these places, including even remote village life, was shaped by the demands of dominant capitalist classes in the economically more developed countries. Large corporations, banks, and governments that served them shaped the world economy, and the now dominant capitalism had a voracious appetite for quick profits.

As the power of the capitalist class shifted from commercial to industrial and then to financial activities, dynamics of capitalist development took on different characteristics. This thesis has been identified with Lenin's theory of imperialism, but, in fact, many other economists, with mainstream, pro-capitalist beliefs, acknowledge the importance of these dynamics even as they may reject the anticapitalist politics associated with some of the critics. Competition, saturated markets, and unplanned imbalances in production leading to flooding of the market all create a downward pressure on the rate of profit. The problem, for certain industries, is too few customers for too many products, and because the economy is so interconnected, a slowdown in one part of the economy often leads to a slowdown in other parts. There may be the same amount of "wealth," money, raw materials, and workers able to work, but if there are not enough customers, the system gradually, or suddenly, shuts down. Metaphorically, it is similar to the medical condition of "shock," where the sudden loss of blood stops the circulation of the remaining blood; there might be enough blood to keep the body alive, but the circulation is not functioning. So too with capital; when the rate of profit drops to the point where enterprises shut down, it has a damaging ripple effect far beyond the initial problem. The system is under pressure to sustain rates of profit.

One way to stabilize profitability is to create customers by extending credit. If one goes into debt to avoid a crisis, and then pays off the debt, this strategy can work. Again, metaphorically, it is comparable to giving a transfusion of plasma or some blood substitute, to at least get the circulation going again. (But if the debt cannot be paid off, it just pushes the crisis back to a later time, when it will be a much bigger crisis. This crisis erupts periodically, most recently in the severe global recession that began in 2007, and again we see the massive creation of artificial spending power through debt being used, this time in the trillions of dollars.) The strategy of extending credit, often associated with Keynes, has certainly been an important part of capitalism for the past eighty years.

A more substantial way to resist this downward pressure on profits is for larger capitalist enterprises to find cheaper labor, cheaper raw materials, and new markets in which to sell their products. The practice of seeking wealth from other lands may be thousands of years old, but the needs of mature capitalism make it a necessity. Acquiring wealth to consume, as was done in precapitalist societies, may be a powerful force for international trade (and conquest), but acquiring profits to keep one's business from failing brings a new level of necessity.

There have been many critics of this thesis, usually quick to point out that the profits made in domestic production far exceeds the profits made through transnational investments. That critique, however, only superficially compares absolute amounts; the important question is whether the profits acquired from global economic activities are necessary for the thriving, or even survival, of these corporations— whatever that absolute amount is. Capitalist economic development can produce immense wealth in a short time. It can also produce great instability in a short time. For example, engaging in agricultural production for profit, rather than for the production of food as such, may result in short-term increases in production, but it also leads to the production of cash crops, which may or may not be edible (or affordable) by the local population. Combine that with business failures that may lead to bankruptcy and the shutting down of small farms, the famines resulting from this process become much larger than those of ancient times.

During this period—mid-1800s until late 1900s—there were huge migrations of people, caused both by the promise of jobs in some areas and the economic and political turmoil in others. The United States accepted tens of millions of immigrants and Western Europe came to rely more and more on immigrant labor as well. The development of large ocean liners and railroads enhanced the ability of people to move, as capital was circulating around the world. Telegraph and then wireless radio accelerated the sharing of information globally. Britain, France, Germany, and the United States sought and gained important profits from international investments. The wars—small wars, medium-sized wars, and especially world wars—carved and carved again "spheres of influence" as the advanced capitalist nations saw their economic competition turn to political rivalries and then military conflict by proxies, eventually leading to full-scale military confrontation, most clearly exemplified by World War I (and to a great extent, World War II). Nevertheless, this was not enough to sustain profit rates and keep the capitalist system in balance. Transnational investments were not enough to stave off capitalism's boom and bust cycles, and the capitalist world was plunged into a worldwide economic depression in the 1930s (Samuelson 2002).

World War II had unintended consequences. Countries were willing to go into considerable debt to finance the war. That debt was used to finance industrial production, which artificially allowed production to expand, mainly through the production of armaments. The deaths of tens of millions of working-age males also alleviated the unemployment crisis. Most importantly, however, the destruction of much of Europe and Japan provided "new markets" and new "customers" for capitalist expansion that had been held back during the 1930s, as these places had to be rebuilt. The destruction of the industrial productive forces during the war meant that overproduction and flooding of the market would be forestalled for a while until capitalist production matured again. This was the basis for what has been called the "postwar economic boom." The political and economic power of Britain, France, and the rest of Western Europe, including, obviously, Germany, were severely weakened, and independence of their former colonies provided more openings for U.S. business to supplant the old colonial interests, although the old colonial powers often did maintain substantial economic investments in many of their former colonies. The same pattern that led to World War I was developing, but the process of carving up the spheres of capitalist investment by the major capitalist powers was maturing more slowly because of the new "market" created by the need to rebuild after World War II.

The growing strength of the Soviet Union in the first decades after World War II was another factor that somewhat delayed the complete spread of U.S.-Western European corporate capitalism to every corner of the world. While one could debate whether the USSR, politically, was on the side of developing socialism as envisioned by Marx, the existence of such a powerful force militarily opposed to U.S. and Western European expansionism did act as a check against unbridled Western capitalist expansion. First, there were large regions of the world that were now allied with the USSR, including China, which were no longer easy places for Western capitalist investment. Second, the existence of the USSR encouraged many nationalist leaders to adopt a "non-aligned" stance to play the West off against the Soviet Bloc as a way of securing a bigger share of the profits for them, and a smaller one for the Western capitalists. Finally, many grassroots movements, often with an anticapitalist thrust, emboldened by the challenge to U.S. power by the Soviets, themselves challenged the power of Western capitalism throughout Asia, Africa, and Latin America in ways that also slowed the acceleration of capitalist expansion around the world (Allison 1989).

During this period, the advances in communication caused some cultural commentators, including Marshall McLuhan, to note the shrinking of the world into a so-called global village (McLuhan 1962). The term "globalization" in an economic context became widely popularized as a result of an article written by Theodore Levitt for the Harvard Business review in 1983 titled "Globalization of Markets," although that word had been used before (Levitt 1983). It was during the following years that globalization accelerated and the term "neoliberal globalization" took root.

Neoliberal Globalization

Neoliberal globalization is an extension of the capitalist-era globalization that predominated from the mid nineteenth century until the late twentieth century. Both neoliberal globalization and the phase that matured in the early twentieth century share similarities with the earlier phase, when capitalism was developing, in that all three of these periods were shaped by the dynamics of capitalist development. However, the latter two periods are fundamentally similar in that they have been driven and shaped by the dynamics of mature capitalism, with the dominance of banks, a more monopolized economy and the needs of major corporations to expand or be swallowed up, and of crises of flooded markets and unstable rates of profit with the concomitant imperative, not just the elective choice, to seek higher rates of profits abroad (Petras and Veltmeyer 2001). While the word "liberal" has come to mean a combination of state-intervention Keynesian economics and "favoring social welfare programs for the poor" in the United States, its meaning in classical economics has been "liberalizing—opening up—the economy for the capitalist class" and minimizing the regulation by the state. As an extension of that, "neoliberal" refers to the resurgence of this "opening up," which intensified in the 1980s as a reaction to the capitalist prosperity and Keynesianism of the previous forty years.

As was discussed earlier, some might see this resurgence of a more unbridled capitalism as a reflection of capitalism's strength in this recent period, while others see it as a superficial, temporary strength caused by the retreat and collapse of the USSR, combined with a fundamental crisis—*a more urgent need* to find new areas in which to invest, especially with secure cheaper sources of labor, as a way to deal with falling rates of profit.

By the early 1970s, the economic boom that accompanied the post-World War II period was already beginning to slow down, as major industrial powers, especially Germany and Japan, had rebuilt their industries and were competing with U.S. capitalism for markets. The once dominant U.S. auto industry faced challenges not just in other countries, but in the United States as well. Even those that were once secondary industrial powers, such as India and Brazil, were flooding the market with goods such as steel. In the United States, there was an initial nationalist thrust toward protectionism. Auto companies such as General Motors, with its not-too-subtle anti-Japanese ad campaign touting Chevrolet as "Hot dogs, apple pie and Chevrolet," and Chrysler, which tried to use Bruce Springsteen's "Born in the USA" as its slogan, were part of a strong, nationalistic "Buy American" campaign. Steel companies joined the bandwagon seeking quotas on imported Japanese steel. The major banks had ambivalent interests—much of their investments were in U.S. corporations threatened by foreign corporations, but they also had major investments in other countries and in sectors that exported.

If one did not trace the economic interests, the policies of various groups might seem very contradictory. For example, in the 1970s, important politically conservative interests were lobbying for the United States to sell more grain to the "communist" Soviet Union while various interests associated with the liberal New York banks, supporters of détente with the USSR, were nevertheless opposed to such sales (Ford 1976). The conservatives were willing to put aside their anticommunist ideology because they saw the opportunity for big profits for their constituencies in the grain-producing states. The mainstream moderates were opposed to the sales, and even went so far as to chastise President Richard Nixon for encouraging them, because they saw these deals causing higher grain prices in the United States, which in turn would fuel inflation that might provoke workers in the major sectors of the economy to demand higher wages, just as the major corporations were trying to cut costs. In more recent times, the archconservative Wal-Mart corporation has no problems dealing with the "communist" government of China. Simply put, the nationalist and internationalist rhetoric of various groups should be seen in the context of their economic interests.

Through the 1970s and into the 1980s, competition among the major advanced capitalist nations intensified as the competitors grew larger and the world grew smaller. This was not absolute, of course. China and India were developing internal markets that somewhat expanded opportunities for investment, but, in general, the drive to sustain profits intensified competition among the advanced capitalist countries to invest worldwide (Berberoglu 2003).

By the late 1980s, both China and the USSR had ceased most of their support for anti-imperialist movements in the Third World, and the nationalist leaders of many of these movements turned away from socialist solutions toward making greater accommodation with the advanced capitalist nations. While it may have intensified the impoverishment of the majority of the people in the Third World, such accommodation allowed for the increase in wealth of the local capitalist class allied with imperialism. In some cases, nationalist leaders who professed support for revolutionary transformation of their societies shifted direction and came, hat in hand, to the World Bank asking for loans, and many of them ended up with considerable wealth while the majority in these countries sank deeper into poverty.

If the developments in the 1970s and 1980s could be seen as the gradual intensification of the same capitalist processes at work before World War I and again after World War II, that of inter-imperialist rivals seeking to sustain their profits by extending their global reach, the 1990s saw another profound development that accelerated capitalist penetration into other parts of the world.

The collapse of the Communist Party regime in the USSR and the privatization of scores of billions of dollars of government-owned assets throughout Eastern Europe provided a boost to West European and U.S. banks and industrial corporations. In many cases, the former Communist Party officials sold the public assets to themselves or to their associates at low prices and then resold them to Western investors. This opened up new markets and new sources of cheap labor and accelerated the globalization of capital. Furthermore, the nationalist leaders of Third World countries were even more inclined to make deals with the World Bank and the International Monetary Fund and encourage foreign investment now that their ability to get aid from the USSR had faded.

In the early 1990s, some economists and social scientists predicted U.S. capitalist hegemony for many years (Fukuyama 2006). Others predicted a triumphant world capitalism, where nation-states would eventually dissolve as transnational corporations facilitated trade and capital flows freely across all borders. In the period around World War I, there was a similar theory, promoted by Karl Kautsky and others, that the nation-state would disappear and a kind of "ultra-imperialism" would become the new world order (Kautsky 1914).

In contrast to this was the viewpoint espoused by Lenin and others that acknowledged the internationalization of capital, but asserted that, ultimately, capitalist groupings would need the political and military power of the nation-state to protect their interests (Semmel 1993). Lenin saw the driving force as being the export of capital, and of investments that capitalist groupings in various nations were compelled to make in order to sustain their profits as markets became saturated domestically and rates of profit were slowing down (Lenin 1969). James Petras and others have argued that, in fact, the term "globalization" is just a softer way of obscuring the intensification of the same forces of modern imperialism that have been at work for the past one hundred years (Petras and Veltmeyer 2001; Petras 2003; Berberoglu 2003; 2005). This modern imperialism sometimes entails direct political rule by a nation over a colony, as France ruled Algeria for much of the twentieth century, but it also is used to fundamentally describe a situation where the "neocolony" has ostensible political independence, but in fact has major parts of its economy controlled by banks and corporations from other countries.

In order for countries to industrialize, there has to be a concentrated accumulation of assets, or, in a capitalist world, an accumulation of capital. This can be acquired through direct conquest and theft, or through state planning, shifting of resources, and so on, or through opening up their economies to foreign investment and loans. The first option is not realistic for the less developed countries. The second option was tried in a number of places but was reversed almost everywhere by the end of the twentieth century. For development, especially industrialization, the third option has been the main option implemented in most of these countries.

Impact of Neoliberal Globalization on the Less Developed Countries

Globalization is promoted as the expansion of "freedom"—by this is meant the freedom to invest and trade through the free flow of capital, goods, and, sometimes, labor without national boundaries or protectionist policies interfering with the "free market." Supposedly, it is a "win-win" situation for the advanced and less developed capitalist countries alike, as they freely bargain in the marketplace. However, the real world is quite different from perfectly balanced equations on an economist's chart. In reality, unregulated economies do not maximize competition in the sense that everyone competes equally. Rather, the unbridled "competition" allows the strongest to set the terms and results in unequal relationships that then build on the advantages/ disadvantages to further intensify the unequal relationships.

In order to secure loans from the World Bank and the International Monetary Fund, the less developed countries have to agree to allow foreign investors to buy up businesses and land in these countries. Furthermore, they have to allow their markets to be opened up to products from the advanced capitalist countries. If there is a surplus of corn grown in the United States, the United States might demand the right to sell that corn at a very low price. This may severely cut into the profits of local farmers, so that many of them will not be able to repay their debts, and therefore will lose their farms—which then are sometimes bought up by transnational agricultural corporations (Stiglitz 2002; 2006). This is one aspect of how globalization works. Ethiopia is the poorest large country on Earth; its per capita Gross Domestic Product income is about US$800 per year in contrast to the United States' $47,000 per year. Its labor force is perhaps the cheapest on Earth, and yet, its shoe industry is in crisis because shoes produced in China are sold for less on the world market, and also in Ethiopia. Ghana's textiles have long been praised for their beautiful designs. In recent years, China has hired local Ghanaians and supplied them with digital cameras. They surreptitiously photograph the latest designs and send the photos over the Internet to Chinese textile enterprises, which then reproduce the designs and sell them back in Ghana at a lower price than the Ghanaian enterprises can match.

Related to this "opening of the economy" are the demands for "economic reform" in the debtor countries. Terms such as "liberalizing the economy" or even "promoting democracy" are catchphrases for demands that the government privatize—that is, sell off to transnational corporations—publicly owned resources (Bello, Cunningham, and Rau 1994). If they refuse to do so, then the loans are in jeopardy. This has resulted, for example, in the sale of public water supplies to private companies, so the local farmers then have to pay for water that once was free. Demands for privatization often results in the selling off of government-run enterprises that are productive and profitable, while local governments still have to run enterprises that function less efficiently, without having the more efficient ones to balance out the losses. These governments are then accused of "inefficiency and waste" because the enterprises they are left to run are losing money. Foreign investment and privatization sometimes does increase the size of upper middle income groups and provide jobs, but for the vast majority of the people in these countries, the standard of living has gone down in the past twenty years, and in some places the average life expectancy has actually dropped as much as ten years or more, despite scientific advances in medicine.

Another aspect of this "economic reform" requires that these governments cut spending as a way to ensure economic stability. Public sector jobs are cut, wages are driven down, and social services are reduced, all of which further lower the standard of living of working people. This increased impoverishment of the working class has also led to another of globalization's consequences: the mass migration of tens of millions of workers from the countryside to the cities and from poor countries to wealthy ones. Albanians risk dangerous sea crossings to get to Italy, East Europeans head to Western Europe or the United States, tens of thousands of South Asians work in the oil fields or do construction work in the Middle East, along with hundreds of thousands of Chinese, who outnumber the native-born population of Dubai, where they work, and significant numbers of Chinese people can be seen in such previously unlikely places as Belgrade, Lagos, or Eastern Ethiopia. In addition to legal immigration, there is also a very large increase in human trafficking—for blue-collar jobs, service jobs, and in the "sex industry."

Finally, neoliberal globalization also creates a kind of "reverse auction" for labor costs, where working people in the United States see their wages driven down as they are forced to compete with cheap labor in the sweatshops of the less developed countries. This has important political implications for the working class in the advanced capitalist countries.

Impact of Neoliberal Globalization on the Advanced Capitalist Countries

In the advanced capitalist countries, some major corporations and banks benefit greatly from neoliberal globalization. Those that can move their production to take advantage of cheaper labor abroad are big winners. Those that can serve as "middlemen," importing products from places where the costs of production are low, are winners. Those that have access to new markets and new customers that were previously unavailable are also winners. On the other hand, those that produce domestically but are forced to compete with those who have access to cheaper labor abroad are losers, as they are often forced out of business or forced into mergers with the winners. Presumably, those who own stock in the winners might be winners. Even for the winners, however, winning is a temporary condition.

In the advanced capitalist countries, most working people are losers from globalization. At some stages of modern imperialism, some sections of the working class might reap immediate financial benefits from imperialism; for example, in the 1950s and 1960s, some U.S. auto companies were reaping such high profits from their factories in Latin America that they were able to grant somewhat bigger wage benefits to the workers in their U.S. plants. Later, though, this wage disparity led to shifting production overseas and a loss of industrial jobs at home, which obviously did not help the workers. It is true that the cost of some of the products that workers buy is held down—electronics from the Pacific Rim countries and clothing produced with cheap labor are two examples. However, as participants in the "reverse auction," many workers see their wages also kept down. As capitalism matures, the trend toward industrialization becomes contradictory in the advanced capitalist countries because of the pressures to cut labor costs that have been increasing in late capitalism. Automation, for example, is both industrialization—as new machines are installed—and, in a sense, deindustrialization, because fewer industrial workers are needed and the middle and upper level blue collar jobs—the path out of poverty for many working-class people—shrinks. Recent globalization has accelerated that process (Bluestone, Cowie, and Heathcott 2003). Jobs are outsourced to other countries, parts of corporations are spun off and sold to other corporations who do not recognize the unions or who move production to parts of the United States (like the South) and the world where production costs are less (Goldman 2005). This, in turn, serves to depress wages in other industries as there is pressure on them to move production to cheap labor areas in the United States or abroad.

While production workers are vulnerable, white-collar workers are perhaps even more vulnerable, as the Internet makes it almost as easy to communicate ideas, develop projects, and send information around the world as it would be to send it to another office in the same building, and inexpensive telephone service makes it much less expensive for Dell Computers to hire an over-the-phone technical support advisor in India than in California—for much lower than half the cost.

Capitalism's contradictory policies toward immigration are another consequence of globalization. The advanced capitalist countries have a long history of turning the immigration flow on and off, depending on their need for cheap labor. However, the past fifteen years have seen mass migrations that are probably unprecedented, certainly at least when compared to any other time in the past ninety years or so. Much of this is not just caused by the needs of the advanced capitalist countries for cheap labor; the impoverished conditions of life have become so severe in many parts of the world that people are forced to emigrate. The governments in less developed countries often encourage this emigration because the money sent home by these workers actually accounts for a significant portion of the income of many of these nations.

The United States does not have a labor shortage, but from the corporations' standpoint, it has a shortage of cheap labor for certain industries, which are the ones supporting more immigration. However, uncontrolled immigration threatens the social order, as capitalism has to strike a balance between labor shortages and labor surpluses that might cause social upheavals. Western Europe, in particular, is changing as a result of massive immigration—they seek immigrants for cheap labor but are uneasy about the possibilities of social disruption, especially because of the immigrants' adherence to the culture of their "home country" and because strong religious and ethno-nationalist attitudes are often strong among new immigrants.

This phase of globalization sets in motion unintended consequences and contradictory outcomes even for the large corporations that initially benefit from this globalization. As a national economy relies more and more on banking and less on domestic industrial production, the potential for wealth increases, but the vulnerability to economic instability also increases. The higher rates of profit are accompanied by the more unstable variables associated with becoming so dependent on overseas investments. Utilizing cheap labor in other countries allows corporations to reap huge profits, but it also lays the basis for rising capitalist groups in other countries to eventually run their own industries and compete with that of the advanced capitalist countries. Furthermore, if, for example, the U.S. corporations deindustrialize at home because they find it more profitable to operate in other countries, or even to just buy from other countries and serve as highly profitable middlemen who do not need to have their capital tied up in material infrastructure assets, the country becomes extremely vulnerable to political developments in other countries and could find itself without the means for needed industrial production in the future.

The strong currency that comes from being an imperialist power results in other countries' choosing to invest in that country—in effect, loaning money to the United States and other countries in similar situations. This brings quick profits to some banks and corporations, and temporary cash flow and prosperity to some segments of the society, but it is prosperity based on debt. If that debt is pulled back, it can lead to higher interest rates, a slowdown of the economy, or even a serious economic depression. The very low rate of savings in the United States, widening gap between the wealthy and especially the middle and lower income working class, the pending crisis in health care costs, the increase in educational costs, and rising debt are all pressures building toward serious problems in the future, as evidenced by the stock market collapse of 2000, which has taken seven years to get back to its previous level, as well as the real estate crisis and the increase in home foreclosures in 2008 and 2009. While it might seem difficult to predict precisely how and when these problems will intensify, it is clear that the so-called boom economy of late capitalism and neoliberal globalization/imperialism bring lots of quick cash to some sectors of the economy, while laying the basis for serious problems for the rest in the future (Brenner 2003).

And, in fact, in late 2007, the United States and the rest of the world experienced another major set of economic problems—the worst since before World War II. When the developing economic slowdown reached a certain point and many loans went into default, credit tightened, and very quickly businesses and private individuals found themselves unable to borrow money that they needed, not just for future purchases but also to pay off their earlier debts. Unable to be sold, property values and stock values dropped. Quite suddenly, the artificially inflated value of real property and of stocks deflated very quickly. Employed people in the United States with private pensions suddenly faced a 40 percent cut in their pension income, and people could not buy new homes because they could not sell their old homes.

What made this a more serious crisis is the reality that the artificially inflated wealth was used to back up other loans that now risked going into default (Bello 2007). The U.S. and many other stock markets lost about 50 percent of their value, home values fell by 20 to 40 percent, and the cash that people and firms used for new purchases disappeared, causing major industries, banking, service and retail businesses to lay off workers, causing an even greater worldwide slowdown. More jobs were lost in the United States in 2008 than in any year since 1945 (*Financial Times*, January 9, 2009). And, according to recent estimates by the U.S. Government, by early 2010 the unemployment rate surpassed the 10 percent mark. These figures, as high as they are, camouflage a greater problem when one considers all the people "off the job market" because of the Iraq and Afghanistan wars, the 2.4 million people incarcerated, and the numbers of part-time and discouraged workers. A major problem is that once the economy slows down beyond a certain point, it cannot easily be boosted back up again. Assuming that there is going to be some respite from this crisis at some point, the realities are that these cycles are endemic to capitalist processes, and they intensify with each new cycle.

To help solve this problem, the U.S. Congress has authorized a "stimulus package" of between 700 billion and several trillion dollars, in hopes that this infusion will get the circulation going again. Besides the obvious question of whether going further into debt is the way to solve a debt problem or just putting it off for the future, there are serious questions about whether this will be enough, given the reality that the actual losses in wealth, whether considered "real" or "artificial," amount to tens of trillions of dollars.

How might this impact international relations? One might assume that the biggest, wealthiest nations will see a need to cooperate to

solve their common problems, and indeed, in the short term, we can
see meetings and conferences designed to encourage cooperation. But
underlying this whole process are serious potential problems as the
advanced capitalist countries compete with each other for profits and
control over the less developed countries (what Lenin called "inter-
imperialist rivalry"), and that can set the stage for sharper conflicts
among the imperialist countries themselves (Lenin 1969). This intense
economic crisis puts even greater strains on these capitalist economies
and pressures them into finding more international sources of profits,
and this, in turn, increases the possibilities for various types of conflicts, not just with smaller countries but with larger ones as well.

World War I appeared to have been started by a conflict between
two different factions from small countries in the Balkans, but these
countries were proxies for the powerful nations that were battling for
much bigger prizes, including Arabian oil. More recently, the U.S.
war in Iraq, begun in 2003, has been characterized by some as a war
for democracy. This has been critiqued by those who point out U.S.
military inaction in the many other areas of the world where the lack
of democracy has hurt many more people. Others see it as a war for
oil. This has been critiqued by those who point out that the United
States has vast quantities of oil, and, in fact, imports very little oil
from Iraq. A more subtle but still economically based analysis sees
the war as largely motivated by the need to control the flow of oil
to Europe, China, and other rivals of U.S. imperialism. Stabilizing
a regime in Iraq that would be friendly to U.S. corporate interests is
seen as providing a military base to protect U.S. oil company interests
in the whole region. It is seen as a way to neutralize Iran, perhaps
turning it into a U.S. ally, as it had been for a part of the twentieth
century. It would protect the profits that U.S. corporations reap as
middlemen, resellers of the region's oil to others (e.g., Europe). It is
not so much the actual oil that the U.S. needs, but rather the huge
profits that are made acquiring and then reselling that oil to others
who need it. Finally, controlling that oil has other important political-economic benefits. Neither France, nor Germany, Japan, Italy,
or Spain own significant sources of oil. Russia has huge amounts of
natural gas, but also eyes the clean, inexpensive Arabian and Iranian
oil. China has growing needs and is fervently seeking new sources of
oil from the Sudan, Eastern Ethiopia, and Nigeria to Venezuela and
Mexico. India, too, will have growing needs. If the U.S. corporations
can maintain tight control on the oil resources of Iraq, and by extension parts of that region, they can maintain an advantage over those

competing oil importers and thus assure U.S. control and domination over the oil resources of the Middle East. It might seem counterintuitive to see allies such as the United States, France, Germany, India, and Japan as rivals to be outmaneuvered by each other, but in a capitalist world, all alliances are ultimately temporary while competition is fundamental. Wallerstein, among others, has argued that there was a sizeable faction within the erstwhile Bush administration that was motivated not just by the so-called Clash of Civilizations between the United States and the radical Islamic movement, but by the economic and political power of Western Europe, Russia, and China as well.

More recently, President Obama has sent a force of over 30,000 more troops to Afghanistan. While Afghanistan may seem to be a poor country with few resources, the reality is that it is strategically located for gas and oil pipelines and for military positioning near Russia, China, and the oil-rich areas in that region.

When the USSR collapsed and much of Eastern Europe pushed aside the various Soviet-style regimes, many mainstream politicians and political theorists postulated that the United States would be the sole superpower for many years to come, the premier world power in a world that was embracing free market capitalism. Even China was opening up its economy to U.S. investments. Within a few years, however, various regional nationalists, especially in the Islamic world, were working to expand their political and economic influence. It was not only the United States that would gain from the collapse of Soviet influence in much of the world. Meanwhile, much of Western Europe moved toward closer economic and political integration, with a unified currency, political alliances, and more coordinated international cooperation on environmental and other policies. This unity might appear to help stabilize the global political situation, but it also creates pressure on some political and economic interests within the United States. The Euro is being used in place of the U.S. dollar in parts of the world, the opposition to U.S. foreign policy, military action, and human rights and environmental policy seems to be growing, and European investments in areas formerly secure for U.S. investments, such as Latin America, are competing with U.S. interests. The European Union, much of which President Bush derided as "Old Europe" in decline, has helped bolster the Hugo Chavez regime in Venezuela and continues to trade with Cuba, as well as lending support to other political movements that are at odds with U.S. imperialism. Currently, the European Union is investing heavily in Mexico. China, too, is rapidly increasing its investments in Latin America. The recent war of words between Russia and the United States, because Russia sees U.S. missiles near its border as a threat, is another example of increased tensions among the great powers. This has been further intensified by the recent conflict between Russia and the former Soviet republic of Georgia, where the United States has been propping up a regime to stir up trouble along the Russian border.

No one is predicting a massive inter-imperialist World War in the near future. The big powers have much to gain from cooperation and much to lose from a major war. However, the increased rivalry among the major capitalist powers in a shrinking world, combined with the rise in economic, technological, and political power of China and India, will create more pressure on all the major capitalist powers. World War I was unthinkable in the early 1890s, the 1917 Bolshevik Revolution and the big influence that the Soviet government had over hundreds of millions of people over the next seventy years was not imagined by anyone twenty years earlier, the rise of defeated Germany to world power status just twenty years after its crushing defeat in World War I was not predicted by many, and the rather sudden collapse of the Soviet Bloc around 1990 and the very different world that has developed since then were also unexpected just twenty years earlier. How the increased economic pressures of today will be resolved cannot easily be predicted, but history should caution us against predicting one hundred years of world peace, especially as today's pressures and crises have become globalized in this shrinking world.

Globalization and Resistance

The poor people of the world, including the working class in most places, seem to be offering little serious resistance to "triumphant" capitalism. The corruption and collapse of the socialist governments and decline in revolutionary movements imply that Marxist egalitarianism as an alternative to capitalism no longer seems to have a significant international movement to give hope and inspiration to the oppressed. There are, however, growing signs of resistance, because desperate people often have no choice but to develop tools and movements to resist. As local groups resist, there is a natural tendency to reach out, across cities, regions, nations, and continents, to seek allies.

It is interesting that Karl Marx and Frederick Engels did not start out *The Communist Manifesto* with a statement about how the capitalists were opposed to communism, but rather with a statement about how the capitalists had declared communism to be dead, although its ghost still frightened them. At various other times, in between attacking Marxism for being materialist and anti-spiritual or for being unrealistic and basically a religion, Marxism has also been declared dead, and the possibility of a world that goes beyond capitalist inequality has been dismissed. Most recently, this has been happening again since the early 1990s.

Within a rather short time, however, the world has seen a resurgence of interest in Marxism, and localized grassroots opposition to global capitalism/imperialism has intensified. The demonstrations against the start of the Iraq War of 2003 were the largest demonstrations in world history. Strikes continue throughout Europe and Latin America, and young people, in particular, are questioning globalization/imperialism. This resistance is nowhere near a revolutionary stage. The resistance is, for the time being, reformist, rather than revolutionary—either in the most narrow economist reforms demanding higher wages, housing, agricultural reform, and so on, or the seemingly most radical reforms, based on ethnic or religious nationalism, demanding political power for one's ethnic or religious group. The resistance generally has a strong anti-U.S. thrust, sometimes also expressing opposition to the economic and political policies of some of the Western European countries. Much of this resistance has a grassroots base, but much of the resistance is currently also under the leadership of local/regional nationalists who seek political-economic power for themselves and their political faction. Some, including Hugo Chavez in Venezuela, Evo Morales in Bolivia, the FARC in Colombia, the New People's Army in the Philippines, the communist-oriented movement in Nepal, and the Zapatista movement in Mexico are secular in their political orientation. Others are using religious concepts to mobilize their base.

The local/regional nationalists who are using religion to mobilize their base pose a particularly difficult problem for the U.S. government. The world economy over the past few decades has left hundreds of millions of people in more desperate poverty than before, and has created deep feelings of alienation and resentment not just among the poor, but also among those who feel solidarity with the poor. The condition of Palestinians on the West Bank and in Gaza remains a sore point for Muslims not just in the Middle East, but throughout the Islamic world, and Israel's recent incursion into Gaza further increases tension between millions of Muslims and the U.S. government. The collapse of an international Marxist movement, which had offered not just allies but provided a worldview that offered a coherent anti-capitalist explanation for the miserable conditions, opened the way for other worldviews that many millions of impoverished and angry people have been seeking in order to make sense out of their condition or the condition of those whom they care for. Religion, combined with nationalism, filled the void and has been skillfully exploited by local leaders.

Ironically, it was the U.S. government and some of its allies that offered considerable aid to some of those groups, seeing them as a bulwark against socialist and communist movements in the 1970s and 1980s. For example, the Israeli government gave considerable support to Hamas in Palestine in the past, and the U.S. government gave substantial support to the Taliban in Afghanistan (Sale 2002). While attention in the United States is often focused on Islamic extremism, similar dynamics are at work in India, where political movements based on an aggressive promotion of the Hindu religion has grown in strength. And while that movement currently sees the United States as an ally against Islamic extremism, it is conceivable that the Hindu nationalists will eventually turn against the United States, especially as India gains in economic and political power. One could even argue that many of the lower income people in the United States currently supporting a movement to give some Christian groups more power in the government actually share many of the same critiques of capitalism as do the Marxists, but the absence of a strong, coherent left movement that can speak to the concerns and alienation of that constituency leaves many of them looking for other worldviews that can explain many of the problems that they face in their daily lives that require solutions.

What are the prospects that either of these tendencies will evolve toward the kinds of revolutionary movements of the past that sought to take the world beyond capitalism and toward a classless society? The religious nationalist movements often use aspects of anticapitalist rhetoric, particularly in denouncing the wealthy and in denouncing the idea that happiness comes from accumulating commodities as dictated by Western capitalism. However, this is combined with an extremely dogmatic reading of religious texts and relies on various religious leaders to be the ultimate interpreters of these texts. Many of the religious leaders affect the demeanor and lifestyle of humble servants, but, in fact, exert very powerful control over their movements. The acquisition of political power and economic wealth seem to accompany their powerful religious authority, despite their antimaterialist rhetoric. The possibility of those movements morphing into secular movements for social change seems remote. Furthermore, some of these local religious-political leaders might eventually ally with one or another major capitalist power in order to protect their interests from other rival capitalist powers. Such developments would serve to sustain capitalist power rather than challenge it.

Some who consider themselves in the anticapitalist, Marxist tradition hail the ascendancy to political power of such reformers as Chavez and Morales, and the World Social Forum has also brought together hundreds of thousands of grassroots activists in conferences where ideas and tactics have been exchanged. It is true that there are aspects of revolutionary Marxism in the mass movements that helped sweep these leaders into power. Throughout Latin America, there is renewed interest in anti-imperialism and strong movements of millions who are developing a critique of modern capitalist imperialism. The prediction that Marxism was dead and that free market capitalism was the "end of history" seems to have been a bit premature. However, while there has been some distribution of wealth, and while some of these leaders strongly criticize the U.S. government, banks, and corporations, none of them are seriously challenging the core of global capitalism. Furthermore, other major capitalist powers, especially from the European Union, as well as China, are making alliances with many of these leaders. While it may seem "progressive" for these leaders to be taking a stance against U.S. imperialism, it is not clear that their leadership will become a major force against capitalism as a world system. Historically, many movements have opposed the dominant imperialist power of the time, only to be co-opted by another, usually rising, capitalist power. In the late 1800s, the United States opposed Spanish imperialism in the Caribbean, for example, and then replaced the Spanish as a rising imperialist power over much of that region. Allying with Western European capitalists against U.S. capitalist interests will not weaken capitalism. China is seen as an eventual competitor of U.S. capitalism, but despite the leadership of the Communist Party, it is clear that capitalist economic relations are growing rapidly in China. Some have suggested that, in the absence of a large, international Marxist movement, China could replace a declining U.S. empire and breathe new life into the world capitalist system.

The grassroots people, the working class and others who are experiencing the worst effects of a global capitalist economy implementing neoliberal policies wherever it can, are the wild cards in this current situation. While the leadership of the various anti-U.S. and anti-West European capitalist movements might not challenge the core of capitalism, there are limits on what severely impoverished people will tolerate. If the coming period sees increased impoverishment, environmental degradation, inadequate responses to natural disasters, and war, we can expect to see major parts of the grassroots people's movements that are currently sustaining these leaders eventually split away and independently organize (Goldman 2005). There are already indications of this, from the land seizures in Latin America to the strikes by oil workers in Iran, workers who are neither tied to the religious politicians nor to the pro-U.S. business reformers. Even in the advanced capitalist countries there is a renewed interest in social change, including internationalism, anti-sweatshop activism, and antiwar organizing. Over one thousand soldiers have signed antiwar petitions in the U.S. military, and many young people are learning how to be organizers all over the world. Imperialism has always provoked rebellion. Neoliberal globalization manifests an intense, accelerated dynamic of exploitation and oppression. Add to this the possibilities of war, or the realities of war, accompanied by a serious economic crisis, and the prospects for mass rebellion can develop quickly. Out of these rebellions can come organizers with a radical perspective, who will look to the roots of social problems to understand these problems and find ways to overcome them. Globalization in the present, specifically neoliberal globalization, might appear to mark the triumph of Western capitalism over its opponents, but the internal contradictions of that system, the limits of that system, the likelihood of rebellion, and the probability that participants in those rebellions will seek to develop theories that explore the roots of problems and actions to uproot the political-economic systems that sustain those problems, all make predictions that "capitalist globalization is the end of history" seem naïve today.

References

Allison, Roy. 1989. *The Soviet Union and the Strategy of Non-Alignment in the Third World*. Cambridge: Cambridge University Press.

Bello, Walden. 2008. A Primer on the Wall Street Meltdown, Transnational Institute, available at http://www.tni.org/detail_page.phtml?act_id=18716.

Bello, Walden, Shea Cunningham, and Bill Rau. 1994. *Dark Victory: The United States, Structural Adjustment, and Global Poverty*. Oakland: California Institute for Food and Development Policy.

Berberoglu, Berch. 2003. *Globalization of Capital and the Nation-State*. Boulder, CO: Rowman and Littlefield.

———. 2005. *Globalization and Change: The Transformation of Global Capitalism*. Lanham, MD: Lexington Books.

Bluestone, Barry, Jefferson Cowie, and Joseph Heathcott. 2003. *Beyond the Ruins: The Meanings of Deindustrialization*. Ithaca, NY: ILR Press.

Brenner, Robert. 2003. *The Boom and the Bubble: The U.S. in the World Economy*. London: Verso.

Diamond, Jared. 2005. *Guns, Germs, and Steel*. New York: W. W. Norton.

Ford, Gerald R. 1976. *Remarks and a Question-and-Answer Session at a Farm Forum in Omaha*. May 7, 1976 (*The American Presidency Project, University of California, Santa Barbara*) http://www.presidency.ucsb.edu/ws/print.php?pid=5958.

Fukayama, Francis. 2006. *The End of History and the Last Man*. New York: The Free Press.

Galeano, Eduardo. 1997. *Open Veins of Latin America: Five Centuries of the Pillage of a Continent*. New York: Monthly Review Press.

Goldman, Michael. 2005. *Imperial Nature: The World Bank and Struggles for Social Justice in the Age of Globalization*. New Haven, CT: Yale University Press.

Harvey, David. 2005. *A Brief History of Neoliberalism*. New York: Oxford University Press.

Kautsky, Karl. 1914. *Ultra-Imperialism*, Die Neue Zeit.

Lenin, V. I. 1969. *Imperialism: The Highest Stage of Capitalism*. New York: International Publishers.

Levitt, Theodore. 1983. "Globalization of Markets." *Harvard Business Review*.

McLuhan, Marshall. 1962. *The Gutenberg Galaxy: The Making of Typographic Man*. Toronto: University of Toronto Press.

Petras, James F. 2003. *The New Development Politics: The Age of Empire and New Social Movements*. Ashgate Publishing.

Petras, James and Henry Veltmeyer. 2001. *Globalization Unmasked: Imperialism in the 21st Century*. London: Zed Books.

Rosen, Ellen Israel. 2002. *Making Sweatshops: The Globalization of the U.S. Apparel Industry*. Berkeley: University of California Press.

Samuelson, Robert J. 2002. "Great Depression." In *The Concise Encyclopedia of Economics*. Indianapolis: Liberty Fund Inc.

Semmel, Bernard. 1993. *The Liberal Ideal and the Demons of Empire: Theories of Imperialism from Adam Smith to Lenin*. Baltimore, MD: The Johns Hopkins University Press.

Stiglitz, Joseph E. 2002. *Globalization and Its Discontents*. New York: W. W. Norton.

———. 2006. *Making Globalization Work*. New York: W. W. Norton.

4

Neoliberalism and the Dynamics of Capitalist Development in Latin America

James Petras and Henry Veltmeyer

An analysis of the dynamics of capitalist development over the past two decades has been overshadowed by an all too prevalent "globalization" discourse. It appears that much of the Left has bought into this discourse, tacitly accepting globalization as an irresistible fact and that in many ways it is progressive, needing only for the corporate agenda to be derailed and an abandonment of neoliberalism. This is certainly the case in Latin America where the Left has focused its concern almost exclusively on the bankruptcy of "neoliberalism," with reference to the agenda pursued and package of policy reforms implemented by virtually every government in the region by the dint of ideology if not the demands of global capital or political opportunism. In this concern, imperialism and capitalism per se, as opposed to neoliberalism, have been pushed off the agenda, and as a result, except for Chavéz's Bolivarian Revolution the project of building socialism has virtually disappeared as an object of theory and practice.

In this chapter we would like to contribute toward turning this around—to resurrect the socialist project; to do so by deconstructing the discourse on "neoliberal globalization" and reconstructing the actual contemporary dynamics of capitalist development.

This is a major task requiring a closer look at the issues. The modest contribution of this discussion is to bring into focus the imperialist dynamics of capitalist development in Latin America. To this end, we present an analytical framework for an analysis of the dynamics of capitalist development and imperialism. We then summarize these dynamics in the Latin American context. Our argument is that the dynamics of capitalist development and imperialism have both an objective-structural and a subjective-political dimension and that a class analysis of these dynamics should include both. This means that it is not enough to establish the workings of capitalism and imperialism in terms of their objectively given conditions that affect people and countries according to their class location in this system. We need to establish the political dynamics of popular and working-class responses to these conditions—to neoliberal policies of structural adjustment to the purported requirements of the new world order. The politics of the Left might so be better informed.

The Neoliberal Era of Capitalist Development and Imperialism

Capitalist development in Latin America can be periodized as follows: (1) an initial phase of primitive accumulation and national development dating more or less from the Independence Movement in the 1820s and crystallizing in the Porfiriato, an extended dictatorship of the big landowners and incipient bourgeoisie in Mexico; (2) a period of modernization, incipient industrialization (in the form of "Fordism") and social reform, dating from the Mexican Revolution in the second decade of the twentieth century; (3) a period of state-led capitalist development with "international cooperation" (technical and financial assistance) dating from the end of World War II and the construction of the Bretton Woods world order (1945–1970); (4) a period of transition (1971–1982) characterized by an extended crisis in the global system of capitalist production and diverse efforts to restructure the system; and (5) the construction of a new world order designed to free the "forces of freedom" from the constraints on capital accumulation imposed by the system of sovereign nation states. This phase, which can be dated from the onset of a region-wide debt and an ensuing "development" crisis, is characterized by dynamic processes of neoliberal globalization and imperialism—the institution of a neoliberal policy framework (the structural adjustment program, as it was termed at the time), a renewed imperial offensive, and the decline, but then partial recovery, of the capital accumulation process and the self-styled "forces of economic and political freedom."

The latest period of capitalist development has two dimensions (globalization in theory/imperialism in practice, forces of opposition and resistance), both of which can also be broken down into four phases.

Neoliberalism and Imperialism in Practice: A Framework of Analysis

Phase I (1975–1982) of the neoliberal project is associated with the bloody Pinochet regime in Chile constituted with a military coup in 1973. The "bold reforms" implemented by this regime and extended into Argentina and Uruguay were subsequently implemented by Margaret Thatcher and Ronald Reagan, and used by economists at the World Bank as a model for the structural reforms set as the price of admission into the new (neoliberal) world order.

Phase II (1983–1990) of neoliberalism (imperialism masked as globalization) includes the foundation stones of a renewed process of capital accumulation on a global scale; setting the parameters for a new configuration of economic and political power; implementation of a second round of neoliberal "structural reform"; launch of an ideology (globalization) designed to legitimate this reform process, and the first wave of privatizations as part of this reform process; and a process of redemocratization designed as a means of securing the political conditions of structural adjustment—a marriage of strategic convenience between capitalism/economic liberalism and democracy/political liberalism (Dominguez and Lowenthal, 1996).

Phase III (1990–2000) entails what might be viewed as a "golden age" of massive transfers of public property to the "private sector" (capitalists and their enterprises); an enormous net outflow of capital ("international resource transfers") in the form of profits on investments, debt payments and royalty charges; virtually no economic growth—less than 1 percent per capita over the decade and a growing divide in the distribution of society's wealth and income; huge bailouts of the banks and investors in corporate stock in a situation of financial crisis; and another round of neoliberal policy reform ("structural reform"), this time with a "human face" (adding to the reform process a "new social policy" targeted at the poor,); a second wave of privatizations and an associated denationalization of the banks and strategic economic enterprises; and a post-Washington Consensus on the need for a more inclusive form of neoliberalism designed to empower the poor (Craig and Porter, 2006; Ocampo, 1998; Van Waeyenberge, 2006).

Phase IV (2000–2009) begins with an involution in the system of capitalist production and the collapse of foreign direct investment inflows, and the onset of political crisis, namely, widespread disenchantment with neoliberalism, and a process of regime change

(Argentina, Bolivia, Ecuador, Brazil, Uruguay, Venezuela—a coup against and the restoration of Chávez to power—and Uruguay). In 2003, the production crisis gave way to a mild economic recovery for a number of countries in the region and a sweeping realignment of political forces into four blocs. The basis of this process of economic and political development was a realignment of global production—a primary commodities boom fueled by the growing demand in China and India for new sources of energy, natural resource industrial inputs, and consumption goods for a rapidly growing middle class.

Opposition to Imperialism, Class Rule, and Neoliberalism: Forces of Resistance

Phase 1 (1973–1982) of the neoliberal project includes a major counteroffensive of the landed proprietors and big capital against the incremental advance of the workers and peasants; a double-offensive of the state against the rural poor and landless peasants in the form of the "Alliance for Progress" ("rural development") and use of the state's repressive apparatus against the guerrilla armies of national liberation, the counteroffensive of capital, with the support of the state, against the working class, resulting in a disarticulation of the labor movement, co-optation of its leadership and a weakening in its capacity to negotiate for higher wages and better working conditions, and, with the agency and support of U.S. imperialism, the institution of military coups and the institution of military rule and a war against "subversives" under the aegis of a Washington-designed "Doctrine of National Security."

Phase II (1983–1999) was characterized by a reorganization of the popular movement, particularly in the countryside—in the indigenous communities and among the masses of dispossessed, landless workers and peasant producers; the mobilization of the forces of popular opposition and resistance against the neoliberal policies of the governments of the day; various uprisings of indigenous peasants in Ecuador, Chiapas, and Bolivia, resulting in the ouster of several presidents if not regime change, and in the blocking of governments efforts to extend the neoliberal agenda; the division of the indigenous movement (in Bolivia and Ecuador) into a social and political movement, allowing it to contest elections as well as mobilize the forces of resistance in direct action against the state; and a general advance in the popular movement with the growth of new offensive and defensive class struggles.

Phase III (2000–2003), corresponding to a crisis in production and ideology vis-à-vis neoliberalism, was characterized by the emergence of various offensive struggles and social mobilizations that led to the overthrow of regimes in Argentina, Bolivia, and Ecuador. In Venezuela, Hugo Chávez came to power, inciting the complex dynamics of a class struggle characterized by a series of counteroffensives by the ruling class (attempted coups, referendums), growing demands for radical reforms, and the institution of the "Bolivarian Revolution" based on an anti-imperialist strategy designed to take the country along a socialist path.

As for Phase IV (2003–2010), it saw the rise of a bloc of pragmatic neoliberal, quasi-populist democratic socialist regimes oriented toward the post-Washington Consensus, an ebb in the flow of the popular movements, the radicalization of Chávez's project of "21st Century Socialism," and the reflux of the popular movement.

Four Cycles of Neoliberalism

"Neoliberalism" in this historic context denotes a national policy—or rather, reform of the then-existing policy of state-led development ("structural reform" or "structural adjustment")—justified with a neoclassical theory of economic growth and development and an ideology of globalization. In this context, we can identify four cycles of neoliberal "structural reform." The first cycle was initiated by the Chicago Boys in Chile under Pinochet. After this first round of neoliberal experiments in policy reform—extended to Argentina and Uruguay, crashed in the early 1980s—a second round of neoliberal policy reforms was implemented under conditions of redemocratization, an external debt crisis, and the political leverage that this crisis provided the World Bank and the IMF, the agencies that assumed primary responsibility for implementing the Washington Consensus on needed policy reform.

The third cycle of neoliberal policies was implemented in the 1990s. At the outset, only four major regimes had failed to fully embrace the "discipline" of structural adjustment. But serious concerns had surfaced as to the sustainability of the neoliberal model and the associated Washington Consensus. For one thing, neoliberalism had utterly failed to deliver on the promise of economic prosperity and mutual benefits to countries North and South of the global development divide. For another, structural reforms had not only released the "forces of freedom" but also the forces of resistance that threatened not only the survival and viability of the neoliberal model but the survival of the state itself. To avert an impending crisis, the ideologues of globalization and neoliberal architects of policy reform came up with a revised model: *structural adjustment with a human face* (UNICEF 1989) in one formulation, *productive transformation with equity* (ECLAC 1990) in another, and "sustainable human development" (UNDP 1996) in yet another. The common feature of these and other such models was a continuing commitment to a neoliberal program of "structural reform" at the level of national policy, the design and adoption of a "new social policy" that "targeted" social investment funds of the poor and their communities, and specific policies that helped shelter the most vulnerable groups from the admittedly high "transitional" social costs of structural adjustment.[1]

Policy Dynamics of Neoliberal Structural Reform

The discourse on "globalization" emerged in the 1980s in the context of efforts in policymaking circles to renovate the ailing Bretton Woods world order—to create a "new world order." Under widespread systemic conditions of a capitalist production crisis and an associated fiscal crisis, economists at the World Bank and its sister "international financial institutions"—all adjuncts of the U.S. imperial state, formulated a program of policy reforms designed to open up the economies of the developing world to the forces of "economic freedom"—to integrate these societies and economies into the new world order. These policy reforms included various IMF stabilization measures such as currency devaluation and import restrictions, and policies of structural adjustment: (1) *privatization* of the means of social production and associated economic enterprises (reverting thereby the nationalization policies of the earlier model of state-led development); (2) *deregulation* of diverse product, capital, and labor markets; (3) *liberalization* of capital flows and trade in products and services; and (4) an *administrative decentralization*, attempting to "democratize" the relation of civil society to the state, transferring to local governments in partnership with civil society the responsibility for economic and social development; that is, privatizing "development" (allowing the poor to "own" and be responsible for improving their lives, changing themselves rather than the system).

By the end of the 1980s, this package of policy reforms had transformed the economic and social system of many Latin American societies. The state-led reforms of the 1960s and 1970s (nationalization, regulation of capitalist enterprise and capital inflows, protection of domestic producers, rural credit schemes, land and income redistribution, market-generated incomes, and so on) had been reverted, effectively halting, where not reversing, the process of development and incremental change.

The outcome and social impacts of this social transformation were all too visible and apparent, especially to those groups and classes that bore the brunt of the adjustment and globalization process. With a significant reduction in the share of labor (and households) in society's wealth and national income, and an equally significant concentration of asset-based incomes and its conversion into capital, Latin American society became increasingly class divided and polarized between a small minority of individuals capacitated and able to appropriate the lion's share of the new wealth and a large mass of producers and workers who had to bear the costs of this "structural adjustment" and who were excluded from its benefits. The economic and political landscape of Latin American society was, and still is, littered with the detritus of this development process. The objectively given conditions of this process are not reflected solely in the all too evident deterioration in living and working conditions of the mass of the urban and rural population. They are also reflected in the evidence of a process of massive outmigration, the export of labor as it were, and an equally massive process of capital export—a net outflow or transfer of "financial resources" estimated by Saxe-Fernandez and Nuñez (2004) to amount to over USD 100 billion for the entire decade of the 1990s. Recent studies suggest that, if anything, the process, fuelled by the financialization of development and policies of privatization, liberalization, and deregulation, has continued to accelerate, putting an end to any talk, and much writing, about a purported "economic recovery" based on a program of "bold reforms" and "sound economics." Neoliberalism is in decline, if not dead.

Globalization or Global Class War?

It is commonplace among many intellectuals, pundits, and policy-makers both in Latin America and elsewhere to discuss "globalization" as if it were a process unfolding with an air of inevitability, the result of forces beyond anyone's control—at worst allowing policy-makers to manage the process and at best to push it in a more ethical direction, that is, allow the presumed benefits of globalization to be spread somewhat more equitably. This is, in fact, the project shared by the antiglobalization movement in their search for "another world," and by the pragmatic center-left politicians currently in power, in their search for "another development."

In this discourse, globalization appears as a behemoth whose appetites must be satisfied and whose thirst must be quenched at all costs—costs borne, as it happens, but not fortuitously, by the working class. In this context, to write, as do so many on the Left today, of the "corporate agenda" and "national interests," and so on, is to obfuscate the class realities of globalization—the existence and machinations of the global ruling class (Petras 2007) and what Jeffrey Faux (2006) terms a "global class war."

Faux's book allows us to view in a different way the globalizing economy, the politics and economics of free trade, and soaring corporate profits on the one hand, and deteriorating standards of living and the continuing (and deepening) poverty of most of the world's people on the other. What is behind this reality? A dynamic objective process, working like the invisible hand of providence through the free market to bring about mutual benefits and general prosperity? Or a class of people who in their collective interest have launched a global war with diverse features and theaters. One feature of this class war, one of many (on its manifestation in the European theater, see Davis, 1984; and Crouch and Pizzorno, 1978) entails ripping up the social contract that had allowed the benefits of capitalism to be broadly shared with other social classes. Another feature was the use of the state apparatus to reduce the share of labor in national income, weaken its organizational and negotiating capacity, and repress any movement for substantive social change.

The globalization discourse hides the class realities behind it. The press, for example, consistently talks about national interests without defining who exactly is getting what and how, under what policy or decision-making conditions. Thus, American workers are told that the Chinese are taking their jobs. But the China threat, in fact, is but another global business partnership, in this case between Chinese commissars who supply cheap labor to global capital and the U.S. and other foreign capitalists who supply the technology and much of the capital used to finance China's exports. Workers in Latin America are told that it is their inflexibility and intransigence, and government interference in the free market, that hold them back from engaging

meaningfully, or engaging at all, in the many benefits of globalization. Many, including those on the Left, view "globalization" in this
way. However, it would be better to see it for what it is: a class project
vis-à-vis the accumulation of capital on a global scale; and as "imperialism" vis-à-vis the project of world domination, a source and means
of ideological hegemony over the system.

Neoliberalism is the reigning ideology of the global elite, a transnational capitalist class that holds its annual meeting in the plush
mountain resort of Davos, Switzerland. Hosted by the multinational
corporations that dominate the world economy (Citigroup, Siemens,
Microsoft, Nestlé, Shell, Chevron, BP Amoco, Repsol-YPF, Texaco,
Occidental, Halliburton, and so on), and attended by some 2000
CEOs, and prominent politicians (including former and the current
presidents of Mexico), this and other such meetings allow this elite to
network with pundits and international bureaucrats, discuss policy
briefs and position papers on the state of the global economy, and
strategize about the world's future—all the while enjoying the best
food, fine wine, good skiing, and cozy evenings by the fire among
friends and associates—with fellow self-appointed and nominated
members and guardians of the imperial world order.

Davos is not a secret cabal, although it is surrounded by meetings
and workings of a host of groupings, that is, meetings and committees
and extended networks. Journalists issue daily reports to the world
on the wit and informal charm of these unelected, self-appointed or
nominated members of the class that run and manage the global economy. In this sense, it is a political convention of what Faux dubs "the
Davos Party," which includes solid representation from the economic
and political elite in Latin America. The mechanism and dynamics
of class membership are unclear. As far as we know it has not been
systemically studied. But it likely involves "people" like Henrique
Fernando Cardoso, former dependency theorist and later neoliberal
president of Brazil, upon or before completion of his term in office,
being invited to give a "talk" or address members of the imperial
brain trust, the global elite, at one of its diverse foundations and "policy forums," such as the Council on Foreign Relations (CFR), a critical linchpin of the imperial brain trust and its system of think tanks,
policy forums, and geopolitical planning centers. Certainly, this is
how former Mexican presidents Carlos Salinas and Ernesto Zedillo
were appointed and assigned specific responsibilities on diverse working "committees" designed to identify and redress fissures in, and
threats to, the system. It is evident that being listed in Forbes' listing

of the world's biggest billionaire family fortunes, such as Bill Gates, George Soros, and Carlos Slim, is sufficient in itself to ensure automatic membership in the club.

The New World Order system easily identifies those members of the global elite in each country that, as Salbuchi (2000) notes, are "malleable, controllable and willing to subordinate themselves to the system's objectives." Their careers are then launched so that they may rise to become presidents of their countries or ministers of finance and central bank governors. This was the case, for example, for Argentina's Domingo Cavallo, Chile's Alejandro Foxley, and Brazil's Henrique Cardoso, each of whom received suitable local and international press coverage; were honored with "prestige-generating" reviews, interviews, conferences, and dinners, and so on; and then invited to address the Council on Foreign Relations, the Americas Society, and Council of the Americas, so that the key New World Order players in New York and Washington could evaluate them. If and when they pass muster, their election campaigns are generously financed by the corporate, banking, and media infrastructure of the "establishment" that has the resources and means to bring them to power legally and democratically—to do the bidding of their masters and colleagues.[2] Some are even invited to join elite circles and organizations (such as Trilateral Commission and the Carnegie Endowment for International Peace), or one of the CFR's working committees.

The Left Responds to the Crisis of Neoliberalism

Throughout the 1990s the dominant popular response to neoliberal globalization and associated regimes and policies was in the form of social movements that represented and advanced most effectively the struggle against neoliberalism and capitalism, in the form of what Ron Chilcote (1990) called a "plurality of resistances to inequality and oppression." These movements placed growing pressure from below on the regime and the "political class." However, by mid-decade, well into the Left's general retreat from class politics, a number of these movements followed Brazil's labor movement (The PT or Workers' party) in establishing a party apparatus to allow them to contest both national and local elections—to pursue an electoral strategy. This political development did not require or mean an abandonment of the social movement strategy of social mobilizations, and so on, but it did open up a broader opportunity to participate in the electoral process, allowing the populace to participate in party politics.

Local Politics and Community Development

The mobilization of the electorate via the institutional trappings of liberal democracy provided a new impetus to the political left—the segment that opted for party politics over social mobilization as a strategy for achieving state power: influencing government policy from within rather than outside the system. However, a large swath of the Left seem to have heeded Jorge Casteñeda's call for the Left to switch its electoral ambitions to the municipality, local politics, and community development. His argument, advanced in *Utopia Unarmed*, was that "municipal politics should be the centrepiece of the Left's democratic agenda...because it typifies the kind of change that is viable...a stepping stone for the future" (1994, 244). Engagement in local politics, he argued—and much of the Left seemed to have followed this line—would provide the basis for a consolidation of the Left after the so-called democratic transition from 1979 (Bolivia, Ecuador) to 1989 (Chile). In addition, it would help rearticulate the civil society-local state nexus and restore legitimacy to the Left's relationship with the popular sector (Lievesley 2005, 8).

An example of the approach proposed by Casteñeda, and, in fact, widely pursued by the Left even before his book (the World Bank's strategy in this regard was already quite advanced) had already been the PT's experience with municipal government in Porto Alegre, the capital city of Brazil's state of Rio Grande do Sul (1989–2004). The PT administration opened up municipal institutions with a stated commitment to accountability and transparency, as well as citizen participation in the budget planning process via the mechanism of public meetings (*Orçamento Participativa*).

The Porto Alegre experience with participatory budgeting was hailed by the World Bank and the International Development "community" of multilateral institutions and liberal academics as a good example of collective decision-making for the common good, a model of grassroots participatory development and politics, and it continues to serve as a guide to similar practices and experiences elsewhere (Abers 1997). Other examples of this "participatory" approach toward local politics and community development, widely adopted by the Left in the 1990s in its retreat from class, can be found in Bolivia and Ecuador. Both countries were laboratories for diverse experiments to convert the municipality into a "productive agent" (the "productive municipality")[3] and exertions by the Left to bring about social change via local politics (North and Cameron 2003). On the left, this shift from macro-politics and development (national elections versus social movements) to micro-politics and development (local politics, participatory development) was viewed as a salutary retreat from a form of analysis and politics whose time had come and gone. Within academe the dynamics of this process has been viewed in some circles as the harbinger of a "new tyranny" (Cooke and Kothari 2001).

The World Social Forum Process: Is Another World Possible?

On January 3, 2007, Caracas, the capital city of an epicenter of social and political transformation in the region, was converted into the Mecca of the international Left. Thousands of activists (10,000 according to the organizers) arrived in Caracas from some 170 countries to participate in the sixth edition of the World Social Forum (WSF), a process initiated in Porto Alegre, Brazil, six years earlier. It was then the first event, which thereafter became an annual event and was extended to and replicated in other regional settings, from India and Europe, to, most recently, Nairobi, Kenya in the African continent. In each place and in each annual event, the organizers would bring together hundreds of nongovernmental and civil organizations committed to the search for a more ethical form of globalization, a more human form of capitalism. The process brings together diverse representatives of a self-defined new left committed to the belief in the necessity and possibility of a "new world," an alternative to globalization in its neoliberal form.

There are, of course, defined limits to this new political process: participants are invited and expected to explore diverse proposals for bringing about "another world," but they are to limit this search to reforms to the existing system, reforms that, no matter how "radical" are expected to leave the pillars of the system intact. This liberal reform orientation to the process is ensured by explicit exclusions— such as those political organizations that include armed struggle or violent confrontation and class struggle in its repertoire, that are oriented toward revolutionary change, and so on.

ATTAC, a Paris-based social democratic organization, is the most visible representative of this approach toward social change, but the World Social Forum from its inception morphed into and became a significant expression of what emerged as the "antiglobalization movement." This movement had its origins in the encounter of diverse forces of resistance formed in middle-class organizations in the "global north" and mounted against the symbols of neoliberal globalization such as the World Trade Organization and the G-7/8 annual summit. A defining moment in this movement, rooted in the organizations of the urban middle class—NGOs, unions, students, and so on—in both Europe and North America, included the successful mobilization against the WTO in Seattle. This mobilization was the first of a number of serialized events scheduled to unfold at important gatherings of the representatives of global capital—Genoa, Quebec, Melbourne, Dakar, and so on.

In Latin America, the World Social Forum process is the basic form taken by the "antiglobalization movement" in the search for "another world" (the latest event in this process was hosted by Lula, taking place in Bélem toward the end of January 2009). Apart from the absence of an internal division between the advocates of moderate reform (ethical globalization) and more radical change, the antiglobalization process is designed to define and maintain the outer limits of permitted change; that is, controlled dissent from the prevailing model of global capitalist development. Not antiglobalization but a more ethical form. Not anticapitalism but a more humane form of capitalism, a more sustainable human form of development. Not anti-imperialism because imperialism is not at issue.

The New Left and the Politics of No-Power

In the shape and form of class struggle the path toward social change in the 1960s and 1970s was paved with state power. That is, the forces of resistance, at the time based in the countryside, in the organizations and movements of the landless and near landless peasants, and in the urban-based organized labor movement—and for the most part led by petit-bourgeois middle-class intellectuals—were concerned with the capture of state power. In the 1990s, in a very different context—neoliberal globalization—and in the wake of the Zapatista uprising in January 1994, there emerged on the Left a postmodern twist to the struggle for social change: "social change without taking state power" (Holloway 2002).

In the discourse of Subcomandante Marcos, the Zapatismo came to symbolically—or theoretically, in the writings of Holloway and others (for example, Burbach 1994)—represent a "new way of doing politics": to bring about social change without resort to class struggle or the quest for state power (Holloway 2002). However, much of the Latin American Left appeared all too ready to retreat from class politics and engage in the new way of "doing politics." Some of the Left joined the struggle for change at the level of local politics and community development—to bring about social change by building on the assets of the poor, their "social capital" (Portes 1998; 2000; Ocampo 2004). Another part joined the "situationists" and other militants of "radical praxis" in an intellectual engagement with the forces of social and political disenchantment in the popular barrios of unemployed workers—in Gran Buenos Aires and elsewhere (Besayag and Sztulwark 2000; Colectivo Situaciónes 2001; 2002). This was in the early years of the new millennium. In the specific conjuncture of economic and political crisis, a generalized rejection of the "old way" of doing politics ("que se vayan todos"), the search for redemption and relevance, left a large part of the Left without a political project, without a social base for their politics.

Dynamics of Electoral Politics: What's Left of the Left

With the advent of the new millennium, it was clear that the neoliberal model, even in its revamped form, had failed to deliver on its promise of economic growth and general prosperity. Instead it had deepened existing class and global divides in wealth and income, and regime after regime was pushed toward its limits of endurance by the forces of popular mobilization. In this context, the political class in each country turned to the Left, opening up new opportunities for groups that had hitherto concentrated their efforts on local politics and community development. Governments of the day, many of them neoliberal client regimes of the United States, fell to the forces of resistance and opposition.

Political developments in the region regarding this regime change led to a concern in the United States, and widespread hopes and expectations on the Left, about a tilt to the left in national politics and what the press (*Globe & Mail*) has termed a "disheartening" triumph of politics over "sound economics." A lot of this concern revolves around Hugo Chávez, who appears (to the press and U.S. policy-makers) to be taking Venezuela down a decidedly anti-U.S., anti-imperialist, and seemingly socialist path—and taking other governments in the region with him.

Chávez's electoral victory was seen by many as the moment when a red tide began to wash over the region's political landscape. In the summer of 2002, the Movement to Socialism (MAS) in Bolivia, led by militant coca growers' leader Evo Morales, became the second largest party in the Congress, while in December it achieved huge victories in municipal elections—in what was billed by the MAS itself as "la toma de los municipios." The election to state power of Lula da Silva in Brazil (October 2002) was followed by Nestor Kirchner in Argentina (May 2003), Tabaré Vasquez in Uruguay (November 2004), Evo Morales (December 2005 and 2006), Rafael Correa in Ecuador (December 2006), and, most recently, Fernando Lugo in Paraguay. The tide was checked in Mexico in the summer of 2006, when Lopez Obrador, presidential candidate of the PRD, fell just short of victory (a well-documented case of electoral fraud), and in Peru, where the nationalist Humala lost out to Alan Garcia, the once disgraced social democrat but reborn neoliberal. But it appeared to swell again with Daniel Ortega's victory in Nicaragua—although, given his opportunism and religious rebirth, Ortega could hardly be viewed as on the Left, notwithstanding his friendship with Chávez and Fidel Castro—and Rafael Correa as well as electoral shifts toward both the Left (Funes in El Salvador and Mujica in Uruguay) and the Right (Pinera in Chile).

Thus, it appeared that Latin America had turned against the U.S.-inspired—and dictated—neoliberal policies of structural adjustment and globalization by electing to state power a number of parties on the political left—although "moderate" or "pragmatic." Center-left regimes, some of which cherish their links with Cuba and relish throwing it in the face of the U.S. administration, which has shown itself to be extraordinarily ideological and non-pragmatic, now out-number right-of-center governments in the region. The days of the U.S.-supported and instigated right-wing dictatorships and military rule are over, having long disappeared in the dustbins of history and replaced by a new breed of neoliberal regimes.

Latin America Turns Left?

These regimes in appearance (that is, as constructed in the rhetoric of public discourse) have changed or are changing economic course, ostensibly moving away from the neoliberal policies pushed by the United States. This was the case in Argentina, for example, where the Kirchner administration was compelled by the most serious economic and political crisis in its history to confront the IMF and the World Bank, and the United States, by halting payments on the country's external debt, redirecting import revenues toward productive and social investments, including short-term work projects demanded by the mass of unemployed workers that at the time constituted over 25 percent of the laborforce, and who had taken to the streets, picketing highways in protest. The result, some three years later, was an annual growth rate of 8 percent, the highest in the region.

Another example of apparent regime change was in Brazil, where in October 2002 the electorate after its third attempt voted Ignacio [Lula] da Silva, leader of the PT, to power, reelecting him in 2006 to a second term in office. Only the second President on the "left" voted into power since Allende in 1970 (after Chávez in 1989), Lula is nevertheless (and for good reason, it turns out) very well received by Wall Street, if not Washington, which tends to view him as a thorn in the U.S. side. Indeed Lula played a major role in defeating the White House plan for a hemispheric free trade zone, and continues to annoy the United States with his support of a Chávez-Morales-Correa axis in Latin American politics. In this context, the intellectual Left associated with the anti-globalization movement chooses to see Lula as an opponent of neoliberal globalization. In fact, Lula, on behalf of Brazil's agribusiness and other capitalist producers, simply has been playing and continues to play hardball in negotiations over access to the U.S. market.

Elections of center-left governments followed in Uruguay (2004), Chile (2005), and Ecuador (2006), where the electorate was polarized between a business magnate, Alvaro Noboa, the richest man in the country and a committed neoliberal ideologue, and Rafael Correa, head of a center-left coalition that appears to be taking Ecuador down the same path as Evo Morales is taking Bolivia, particularly in regard to a constituent assembly that might well, or is expected to, change the economic and social system as well as the correlation of class forces in the country's politics. In this regard, elements of the political Left in Ecuador, especially those associated with the "Coordinadora de Movimientos Sociales" (CMS), see a political opportunity to build a "radical bloc" on the basis of combined action "from above" (the government) and "from below" (the indigenous and popular movement). Whether this will happen (see Saltos 2006)[4] remains to be seen. For one thing, it hinges on the capacity of the popular movement for active mobilization—to pressure the Correa government from below toward the left. On this, the historic record is fairly clear. As observed by Pedro Stedile, leader of the MST, "without active mobilization the government gives nothing."

With the election of Rafael Correa over Alvaro Noboa, the popular and indigenous movement in Ecuador at least placed on the agenda of government action issues such as national sovereignty, nationalization of the country's natural resources, agrarian reform, indigenous rights, subordination of payment on the external debt to social programs, renegotiation of oil contracts with the multinationals, the ending of the military bases in Manta, and Latin American (vs. continental) integration. Whether the government will act on these issues remains to be seen.

The conflict that ensued over the Constituent Assembly (CA) in Ecuador and Bolivia, where the CA was finally approved in January 2009, is symptomatic of the profound legitimation crisis in the system of class domination in these and other countries (Saltos 2006). Other earlier forms of hegemony, such as "globalization" and the trappings of representative "democracy," have lost their hold over people, having been totally undermined by the all too tangible and visible signs of the negative effects of neoliberal policies. The reign of Washington in the region appears to be in serious decline. Nor can Washington, in its efforts to preserve the status quo or the status quo ante, revert to the use of force—to bring back the Armed Forces to restore order. Its only recourse is to engage "civil society" in the project of "good governance"—to restore political order by means of a broad social consensus that reaches well beyond the state and the political class (Blair 1997; OECD, 1997; UNDP, 1996; World Bank, 1994b).

What we saw in Quito and La Paz in regard to the Constituent Assembly went beyond a conflict between two branches of government. At issue was that those who elected Correa and Morales had come to the point of refusing to be subordinated to a state controlled by the dominant class and servile to Washington and the interests of global capital. On achieving political representation with the election of Morales and Correa, and Chávez, for that matter, the forces in the popular movement were all too aware that the legislature was dominated by the "oligarchy" (as the ruling class is understood in Bolivia and Ecuador). In this situation, Morales and Correa were compelled to construct a multi-class alliance and mobilize the forces of resistance to class rule and the neoliberal agenda of previous governments under the post-Washington Consensus. The result is the construction of a multiethnic or plurinational state oriented toward what the vice president of Bolivia, Alvaro Garcia, conceives of as an Andean form of capitalism, and a new anti-American axis of regional politics and trade.

These and other such political developments in Bolivia and Ecuador are illustrative of what appears to be a regional trend. For example, in neighboring Colombia in October 2003 the voters elected a former union leader, Luis Garzón, as mayor of Bogotá. The election marked a swing to the Left in Colombia's second most important elective office, a clear challenge to the pro-U.S., scandal-ridden, right-wing government of Alvaro Uribe. If we take these and other such developments together, especially in Venezuela, Bolivia, and Ecuador, there does indeed seem to be a leftward swing in the political winds of change.

Whither Socialism in a Sea of Crisis and Neoliberal Decline?

A serious discussion of the prospects for socialism in Latin America today must take into account world economic conditions in the current conjuncture, the state of U.S.-Latin American relations relative to the project of world domination and imperialism, the specific impact on Latin American countries of these conditions and relations, the conditions deriving from the correlation of class forces within these countries, and the class nature and agency of the state relative to these forces.

World Economic Conditions and Their Impact on Latin America

Latin America's "restructured" capitalist economy emerged from the financial crisis of the 1990s and the recession of the early years of the new millennium with its axis of growth anchored in the primary sector of agro-mineral exports (Cypher 2007; Ocampo 2007). From 2003 to 2008, all Latin American economies, regardless of their ideological orientation or political complexion, based their economic growth strategy on the "re-primarization" of their export production, to take advantage thereby of the expanding markets for oil, energy, and natural resources and the general increase in the price of primary commodities on the world market. The driving force of capitalist development in this period was agribusiness and mineral exports, export-oriented production of primary commodities leading to an increased dependence on diversified overseas markets and a change in the correlation of class forces, strengthening the Right and, notwithstanding a generalized tilt to the left at the level of the state, a weakening of the Left. Ironically, the primarization of exports led to the revival and strengthening of neoliberalism via the reconfiguration of state policy to favor agro-mineral exporters and accommodate the poorest section through populist clientelistic "poverty programs." In the context of a primary commodities boom and the emergence of a range of democratically elected center-left regimes, trade union leaders were co-opted and the social movements that had mobilized the forces of resistance to neoliberalism in the 1990s were forced to beat a retreat from the class struggle (Petras and Veltmeyer 2009).

The link between U.S. finance capital, the growth of industry and the domestic market in Asia, and the primary commodities boom was responsible for the period of high growth in Latin America from 2003 to 2008, when the boom went bust and most economies in the region succumbed to a financial crisis of global proportions and a system-wide deep recession that threatened to push the U.S. economy, at the center of the gravitational force of this crisis, toward collapse. With the U.S. empire's "over-extension" and the exceedingly high costs of prosecuting imperialist war in Iraq and maintaining its enormous military apparatus—military expenditures on the Iraq war alone increasing by millions each minute (as of February 17, 2009 US$ 597.7 billion) and likely to cost well over a trillion dollars before it is over—the capacity of the United States to weather the storm of financial crisis and a deepening recession has been seriously diminished. Given the absorption of the U.S. state in the Iraq war, governments in Latin America in the latest phase of capitalist development managed to achieve a measure of "independence" and "relative autonomy" in their relations with the United States. And this has given leaders like Hugo Chávez a free hand in his efforts to push Venezuela in a social-ist direction.

Impact of World Recession and U.S. Imperial Revivalism in Latin America

Latin America is feeling the full brunt of world recession. Every coun-try in the region, without exception, is experiencing a major decline in trade, domestic production, investment, employment, state rev-enues, and income. The projected growth of Latin America's GDP in 2009 has declined from 3.6 percent in September 2008 to 1.4 percent in December 2008 (*Financial Times*, January 9, 2009). More recent projections estimate Latin America's GDP per capita as falling to minus 2 percent.[5] As a result, state spending on social services will undoubtedly be reduced. State credit and subsidies to big banks and businesses will increase, and unemployment will expand, especially in the agro-mineral and transport (automobile) export sectors. Public employees will be retrenched and will experience a sharp decline in salaries. Latin America's balance of payments will deteriorate as the inflow of billions of dollars and euros in remittances from over-seas workers, a major source of "international financial resource" for many countries in the region, declines. Foreign speculators are already withdrawing tens of billions of investment dollars to cover their losses in the United States and Europe. A process of foreign disinvestment has replaced the substantial inflow of "foreign investment" in recent years, eliminating a major source of financing for major "joint ventures." The precipitous decline in commodity prices in 2008, reflecting an abrupt drop in world demand, has sharply reduced government revenues dependent on export taxes. Foreign reserves in Latin America can only cushion the fall in export revenues for a limited time and to a limited extent.

The recession also means that the economic and social structure, the entire socioeconomic class configuration on which Latin America's growth dynamic in recent years (2003–2008) was based, is headed for a major transformation. The entire spectrum of political parties linked to the primary commodity export model and that dominate the electoral process will be adversely affected. The trade unions and social movements oriented toward an improvement in their socioeconomic conditions and wages, social reforms, and increased expenditures of fiscal resources and social spending within the primary commodity export model will be forced to take direct action or lose influence and relevance.

The initial response of the left-of-center regimes that came to power in the context of a primary commodities boom and the demise of neoliberalism has largely focused on (1) financial support for the banking sector (Lula) and lower taxes for the agro-mineral export elite (Kirchner/Lula); (2) cheap credit for consumers to stimulate domestic consumption (Kirchner); and (3) temporary unemployment benefits for workers laid off from closed small and medium size mines (Morales). The response of the Latin American regimes to date (up to the beginning of 2009) could be characterized as delusional, the belief that their economies would not be affected. This response was followed by an attempt to minimize the crisis, with the claim that the recession would not be severe and that most countries would experience a rapid recovery in "late 2009." It is argued in this context that the existing foreign reserves would protect their countries from a more severe decline.

According to the IMF, 40 percent of Latin America's financial wealth ($2.2 billion) was lost in 2008 because of the decline of the stock market and other asset markets and currency depreciation. This decline is estimated to reduce domestic spending by 5 percent in 2009. The terms of trade for Latin America have deteriorated sharply as commodity prices have fallen sharply, making imports more expensive and raising the specter of growing trade deficits (*Financial Times*, January 9, 2009, 7).

The impact of these "developments" can be traced out not only in regime politics but on the class structure and the correlation of forces associated with this structure. Thus, the fall in the demand and price of primary commodities is resulting in a sharp decline in income, the power and the solvency of the agromineral exporters that dominated state policy in recent years. Much of their expansion during the "boom years" was debt-financed, in some cases with dollar- and euro-denominated loans (*Financial Times*, January 9, 2009, 7). But many of the highly indebted "export elite" now face bankruptcy and are pressuring their governments to relieve them of immediate debt obligations. And in the course of the recession/depression there will be a further concentration and centralization of agro-mineral capital as many medium and large miners and capitalist farmers will have to foreclose or will be forced to sell. The relative decline of the contribution of the agro-mineral sector to the GDP and state revenues means they will have less leverage over the government and economic decision-making. The collapse of their overseas markets and their dependence on the state to subsidize their debts and intervene in the market mean that the "neoliberal" free market ideology is dead—for the duration of the recession. Weakened economically, the agro-mineral elite are turning to the state as its instrument of survival, recovery, and refinancing.

In this new context, the "new statism" in formation has little or nothing "progressive" about it, let alone any claim to "socialism." The state under the influence of the primary sector elites assumes the primary task of imposing the entire burden of the recession on the backs of the workers, employees, small farmers, and business operators. In other words, the state is charged with indebting the mass of people in order to subsidize the debts of the elite export sector and provide zero cost loans to capital. Massive cuts in social services (health, pensions, and education) and salaries will be backed by state repression. In the final analysis the increased role of the state will be primarily directed to financing the debt and subsidizing loans to the ruling class.

The State of U.S. Relations in Latin America in the Current Conjuncture

If the United States suffered a severe loss of influence in the first half decade of the early 2000s due to mass mobilization and popular movements ousting its clients, during the subsequent four years the United States retained political influence among the most reactionary regimes in the region, especially in Mexico, Peru, and Colombia. Despite the decline of mass mobilizations after 2004, the aftereffects continued to ripple through regional relations and *blocked* efforts by Washington to return to relations that had existed during the "golden decade" of pillage (1990–1999).

While internal political dynamics put the brakes on any return to the 1990s, several other factors undermined Washington's assertion of full-scale dominance: The United States turned all of its attention, resources, and military efforts toward multiple wars in South Asia (Afghanistan), Iraq, and Somalia and to war preparations against Iran while backing Israel's aggression against Palestine, Lebanon, and Syria. Because of the prolonged and losing character of these wars, Washington remained relatively immobilized as far as South America was concerned. Equally important, Washington's declaration of an intensified worldwide counterinsurgency offensive (the "War on Terror") diverted resources toward other regions. With the U.S. empire builders occupied elsewhere, Latin America was relatively free to pursue a more autonomous political agenda, including greater regional integrations, to the point of rejecting the U.S. proposed "Free Trade Agreement."

In this new context the spectrum of international relations between the United States and Latin America runs the gamut from "independence" (Venezuela), "relative autonomy" within competitive capitalism (Brazil), relative autonomy and critical opposition (Bolivia), to selective collaboration (Chile) and deep collaboration within a neoliberal framework (Mexico, Peru, and Colombia). Venezuela constructed its leadership of the alternative nationalist pole in Latin America in reaction to U.S. intervention. The Chávez government has sustained its independent position through nationalist social welfare measures, which has garnered mass support. A policy of "independence" was made possible, and financed as it were, by the commodity boom and the jump in oil prices. The "dialectic" of the U.S.-Venezuelan conflict evolved in the context of U.S. economic weakness and overextended warfare in the Middle East, on the one hand, and economic prosperity in Venezuela, which allowed it to gain regional and even international allies, on the other.

The autonomous-competitive tendency in Latin America is embodied by Brazil. Aided by the expansive agro-mineral export boom, Brazil projected itself on the world trade and investment scene, while deepening its economic expansion among its smaller and weaker neighbors such as Paraguay, Bolivia, Uruguay, and Ecuador. Brazil, like the other BRIC countries, which include Russia, India, and China, forms part of a newly emerging expansionist power center intent on competing and sharing with the United States the control over the region's abundant resources and the smaller countries in Latin America. Brazil under Lula shares Washington's economic imperial vision (backed by its armed forces), even as it competes with the United States for supremacy. In this context, Brazil seeks extra-regional imperial allies in Europe (mainly France) and it uses the "regional" forums and bilateral agreements with the nationalist regimes to "balance" its powerful economic links with Euro-U.S. financial and multinational capital.

At the opposite end of the spectrum are the "imperial collaborator" regimes of Colombia, Mexico, and Peru, which remain steadfast in their pro-imperial loyalties. They are Washington's reliable supporters against the nationalist Chávez government and staunch backers of bilateral free trade agreements with the United States.

The other countries in the region, including Chile and Argentina, continue to oscillate and improvise their policies in relation to and among these three blocs. But what should be absolutely clear is that all the countries, whether radical nationalist or imperial collaborators, operate within a capitalist economy and class system in which market relations and the capitalist classes are still the central players.

Socialism and the Latin American State in the Current Conjuncture of the Class Struggle

Control of the state is an essential condition for establishing socialism. But it is evident that a more critical factor is the composition of the social forces that have managed to achieve state power by one means or the other. From 2003 to 2008, in the context of a primary commodities boom and a serious decline in the mobilizing power of neoliberal globalization, one state after the other in Latin America has tilted to the Left in establishing a nominally anti-neoliberal regime. However, the only regime in the region with a socialist project is that of Chávez, who has used the additional fiscal resources derived from the sale of oil and the primary commodities boom—specifically the growing world demand for oil—to turn the state in a socialist direction under the ideological banner of the "Bolivarian Revolution." All of the other center-left regimes formed in this conjuncture for one reason or the other, and regardless of their national sovereignty concerns vis-à-vis U.S. imperialism, have retained an essential commitment to neoliberalism, albeit in a more socially inclusive and pragmatic form as prescribed by the post-Washington Consensus (Ocampo 1998). A surprising feature of these center-Left regimes is that not one of them—again Venezuela (and, of course, Cuba) being the exception—use their additional fiscal revenues derived from the primary commodities boom to reorient the state in a socialist direction, that is, to share the wealth or, at least, in the absence of any attempt to flatten or eliminate the class structure, to redirect fiscal revenues toward programs designed to improve the lot of the subordinate classes and the poor. Again, Chávez is the exception in the use of windfall fiscal revenues derived from the primary commodities boom (oil revenues in the case of Venezuela) to improve conditions for the working class and the popular classes. The statistics regarding this "development" (see Weisbrot 2009) are startling. Over the entire decade of Chávez's rule, social spending per capita has tripled and the number of social security beneficiaries more than doubled; the percentage of households in poverty has been reduced by 39 percent, and extreme poverty by more than half. During the primary commodities boom (2003–2008), the poverty rate in Venezuela was cut by more than half, from 54 percent of households in the first half of 2003 to 26 percent at the end of 2008. Extreme poverty fell even more (by 72 percent). And these poverty rates measure only cash income, and do not take into account increased access to health care or education. However, in the other countries in the region governed by left-leaning regimes, not one of which is oriented toward socialism, conditions were and are very different. In a few cases (Chile, Brazil) the rate of extreme poverty was cut, but in all such cases, despite recourse to an antipoverty program following the PWC, government spending was relatively regressive. In only one case (Venezuela) are per capita fiscal expenditures greater today than it was in 2000 in the vortex of a widespread crisis and a zero growth (Clements, Faircloth, and Verhoeven 2007). In many cases, social programs and government spending was allocated so as to distribute more benefits to the richest stratum of households and the well-to-do than to the working class and the poor.[6] Even in the case of Bolivia, where the Morales-Garcia Linera regime has a clearly defined anti-neoliberal and anti-U.S. imperialist orientation, not only has the government not expanded social program expenditures relative to investments and expenditures designed to alleviate the concerns of foreign investors, but the richest stratum of households benefited more from fiscal expenditures on social programs than the poorest (Petras and Veltmeyer 2009). All of the center-left regimes that came to power in this millennium, especially Brazil and Chile, elaborated antipoverty programs with reference to the PWC. In the case of Bolivia, fiscal expenditures on social programs defined by the "new social policy" of the post-Washington Consensus have been supplemented by a populist program of bonuses and handouts, and popular programs in health and education, but these have been almost entirely financed by Cuba and Venezuela. As for the fiscal resources derived from Bolivia's participation in the primary commodities boom, they have been allocated with a greater sensitivity to the concerns of foreign investors than the demands of the working class and the indigenous poor.

In this situation, what is needed is not only access to state power, which the social movements managed to ostensibly achieve via the election of Evo Morales, but an ideological commitment of the government to socialism—to turn the state in a socialist direction. In this connection the Chávez regime is unique among Latin American regimes. Even so, the road ahead for the Bolivarian revolution in bringing about socialism of the twenty-first century promises to be long and "rocky," as in the case of Cuba, littered with numerous pitfalls, but, unlike Cuba, with the likely growth in the forces of opposition.

Notes

1. The basic elements of the new post-Washington Consensus policy agenda under the model of "sustainable human development (UNDP, 1996) are: (1) a neoliberal program of macroeconomic policy measures, including privatization, agricultural modernization, and labor reform; (2) a "new social policy" supported by a "social investment fund" targeted at the poor; (3) specific social programs (policies related to health, education, and employment) designed to protect the most vulnerable social groups from the brunt of the high "transitional" social costs of structural adjustment—and to provide a "human face" to the overall process; and (4) a policy of administrative decentralization and popular participation designed to establish the juridical-administrative framework for a process of participatory development and conditions of "democratic governance."

2. Of course, this also applies to the United States, as in the run-up to George W. Bush's campaign for a second term in office. On July 28, 2004, a caravan of fifty multibillionaires met in Boston to defend and secure the electoral victory of the president. In the words of Count Mamoni—to a reporter of *La Jornada* (Jul 28, 2004), "We are the rich who wish to ensure that the president who we bought [paid for] stays in the White House." He adds that "those of us who were born to wealth and privilege...[are] owners of the country [and must

continue as such]." One of the participants in the "Join the Limousine" tour added that "we are all winners under this government, just some a lot more than others."

3. On this, see De la Fuente (2001), Sánchez (2003), and Terceros and Zambrana Barrios (2002).

4. Napoleon Saltos, Director of the CMS, sees political developments in Ecuador as somewhere between Venezuela, which is implementing from above a sort of socialist plan without pressure from below, and Bolivia, where the government, to some extent, is subject to the pressures of a mobilized population.

5. The onset of the recession in Latin America is evident in the 6.2 percent fall in Brazil's industrial output in November 2008 and its accelerating negative momentum (*Financial Times*, January 7, 2009, 5).

6. On this point, see the IMF as in Alier and Clements (2007: 4–5): "Reallocating social spending to programs that most benefit the poor... [are] important for forging a more equitable society... [but] the distributive incidence of social spending varies greatly across programs, with primary education and social assistance programs having the most favorable impact, while higher education and social insurance programs tend to benefit middle and upper-income groups. Because of the low share of spending in pro-poor programs—such as social assistance—the majority of social spending benefits accrue to those that are relatively well off."

References

Abers, Rebecca. 1997. *Inventing Local Democracy: Neighborhood Organizing and Participatory Policy-Making in Porto Alegre, Brazil*. Ph.D. Dissertation, University of California, Urban Planning.

Alier, Max and Benedict Clements. 2007. "Comments on Fiscal Policy Reform in Latin America," Paper prepared for the Copenhagen Consensus for Latin America and the Caribbean—*Consulta de San José*, Costa Rica, October 20–25,

Aznar, José María. 2007. *América Latina. Una agenda de libertad*. Madrid: Fundación para el Análisis y los Estudios Sociales (FAES).

Besayag, Miguel y Diego Sztulwark. 2000. *Política y situación: de la potencia wl contrapoder*. Buenos Aires: Ed. De Mano en Mano.

Blair, H. 1997. *Democratic Local Governance in Bolivia*. CDIE Impact Evaluation, No. 3. Washington, DC: USAID.

Booth, David. 1996. "Popular Participation, Democracy, the State in Rural Bolivia," Dept. of Anthropology, Stockholm University. La Paz.

Bulmer-Thomas, Victor. 1996. *The New Economic Model in Latin America and its Impact on Income Distribution and Poverty*. New York: St. Martin's Press.

Burbach, Roger. 1994. "Roots of the Postmodern Rebellion in Chiapas," *New Left Review*, 1, 205.

Casteñeda, J. G. 1994. *Utopia Unarmed: The Latin American Left After the Cold War*. New York: Vintage.

Chilcote, R. H. 1990. "Post-Marxism. The Retreat from Class in Latin America," *Latin American Perspectives* 65, 17, Spring.

Clements, Benedict, Christopher Faircloth and Marijn Verhoeven. 2007. "Public Expenditure in Latin America: Trends and Key Policy Issues," *IMF Working Paper* WP/07/21.

Colectivo Situaciones. 2001. *Contrapoder: una introducción.* Buenos Aires: Ediciones de Mano en Mano, Noviembre.

———. 2002. *19 y 20: Apuntes para el nuevo protagonismo social.* Buenos Aires: Editorial De mano en Mano, Abril.

CONAIE. 1994. *Proyecto político de la CONAIE.* Quito.

CONAIE—Confederación de Nacionalidades Indígenas de Ecuador. 2003. *Mandato de la I Cumbre de las Nacionalidades, Pueblos y Autoridades Alternativas.* Quito: CONAIE.

Cooke, B. and U. Kothari, eds., 2001. *Participation: The New Tyranny?* London and New York: Zed Books.

Crabtree, John. 2003. "The Impact of Neo-Liberal Economics on Peruvian Peasant Agriculture in the 1990s," in Tom Brass, ed., *Latin American Peasants,* 131–161. London: Frank

Craig, D. and Porter, D. 2006. *Development Beyond Neoliberalism? Governance, Poverty Reduction and Political Economy.* Abingdon, Oxon: Routledge.

Crouch, C. and Pizzorno, A. 1978. *Resurgence of Class Conflict in Western Europe Since 1968.* London: Holmes & Meier.

Cypher, James M. 2007. "Back to the 19th Century? The Current Commodities Boom and the Primarization Process in Latin America," Presented to the LASA XXVII International Congress Session ECO20, Montreal, Canada, September 5–8.

Dávalos, Pablo. 2004. "Movimiento indígena, democracia, Estado y plurinacionalidad en Ecuador," *Revista Venezolana de Economía y Ciencias Sociales* 10, 1, Enero–Abril.

Davis, Mike. 1984. "The Political Economy of late-Imperial America," *New Left Review,* 143, January–February.

———. 2006. *Planet of Slums.* London: Verso.

De Castro Silva, Claudete y Tania Margarete Keinart. 1996. "Globalizacion, Estado nacional e instancias locales de poder en America Latina," *Nueva Sociedad,* 142, Abil-Mayo.

De la Fuente, Manuel, ed. 2001. *Participación popular y desarrollo local,* Cochabamba: PROMEC-CEPLAG-CESU.

De la Garza, Enrique. 1994. "Los sindicatos en America Latina frente a la estructuración productiva y los ajustes neoliberales," *El Cotidiano,* No. 64, 9–10, Mexico.

Delgado-Wise, Raúl. 2006. "Migration and Imperialism: The Mexican Workforce in the Context of NAFTA," *Latin American Perspectives,* 33 (2): 33–45.

Dominguez, J. and A. Lowenthal, eds. 1996. *Constructing Democratic Governance.* Baltimore: John Hopkins University Press.

ECLAC—Economic Commission for Latin America and the Caribbean. 1990. *Productive Transformation with Equity.* Santiago, Chile: ECLAC.

Faux, Jeffrey. 2006. *The Class War.* Washington DC: Economic Policy Institute.

Holloway, John. 2002. *Change The World Without Taking Power: The Meaning of Revolution Today*. London: Pluto Press.

Holloway, John and Eloina Peláez, eds. 1998. *Zapatista! Reinventing Revolution in Mexico*. London: Pluto Press.

Levitt, Kari. 2003. "Grounding the Globalization Debate in Political Economy," Notes for a Contribution Toward the publication of *Globalization and Anti-Globalization*. Halifax: Saint Mary's University.

Lievesley, Geraldine. 2005. "The Latin American Left: The Difficult Relationship between Electoral Ambition and Popular Empowerment," *Contemporary Politics* 11, 1, March.

Macas, Luis. 2000. "Movimiento indígena ecuatoriano: Una evaluación necesaria," *Boletín ICCI "RIMAY,"* Año 3, 21, diciembre, 1–5.

———. 2004. "El movimiento Indígena: Aproximaciones a la comprensión del desarrollo ideológico politico," *Tendencia Revista Ideológico Político*, I, Quito, Marzo, 60–67.

Marcos, Subcomadante. 1994. "Tourist Guide to Chiapas," *Monthly Review*.

North, Liisa and John Cameron, eds. 2003. *Rural Progress, Rural Decay: Neoliberal Adjustment Policies and Local Initiatives*. Bloomfield, CT: Kumarian Press.

Ocampo, A. 2004. "Social Capital and the Development Agenda," in Atria, R. ed., *Social Capital and Poverty Reduction in Latin America and the Caribbean: Towards a New Paradigm*, 25–32. Santiago: United Nations.

Ocampo, José Antonio. 1998. "Beyond the Washington Consensus: An ECLAC Perspective," *CEPAL Review* 66, December: 7–28.

———. 2007. "The Macroeconomics of the Latin American Economic Boom," *CEPAL Review* 93, December.

OECD—Organisation of Economic Cooperation and Development. 1997. *Final Report of the DAC Ad Hoc Working Group on Participatory Development and Good Governance*. Paris.

Petras, James. 1997a. "The Resurgence of the Left," *New Left Review*, 223.

———. 1997b. "MST and Latin America: The Revival of the Peasantry as a Revolutionary Force," *Canadian Dimension* 31, 3, May/June.

———. 2001. "Are Latin American Peasant Movements Still a Force for Change? Some New Paradigms revisited," *The Journal of Peasant Studies* 28, 2.

———. 2006. "Following the Profits and Escaping the Debts: International Immigration and Imperial-Centered Accumulation." *Dissident Voice*, August 8 http://dissidentvoice.org/Aug06/Petras08.htm.

———. 2007. "Global Ruling Class: Billionaires and How They 'Made It,'" *Global Research*, March 23 [www.globalresearch.ca]

Petras, James and Henry Veltmeyer. 2005. *Social Movements and the State: Argentina, Bolivia, Brazil, Ecuador*. London: Pluto Press.

———. 2009. *What's Left in Latin America*. Aldershot: Ashgate Publishing.

Portes, A. 1998. "Social Capital: its Origins and Applications in Modern Sociology," *Annual Review of Sociology*, 24: 1–24.

———. 2000. "Social Capital: Promise and Pitfalls of its Role in Development," *Journal of Latin American Studies*, 32: 529–547.

Salbuchi, Adrian. 2000. *El cerebro del mundo: la cara oculta de la globalización.4th. ed.*, Córdoba, Argentina: Ediciones del Copista.

Saltos Galarza, Napoleón. 2006. "La derrota del poder económico y la emergencia del poder constituyente," Quito, December 1. wnsaltosg@yahoo.es.

Sánchez, Rolando, ed. 2003. *Desarrollo pensado desde los municipios: capital social y despliegue de potencialidades local.* La Paz: PIED—Programa de Investigación Estratégia en Bolivia.

Saxe-Fernández, John and Omar Núñez. 2001. "Globalización e Imperialismo: La transferencia de Excedentes de América Latina," in J. Saxe-Fernández, J. Petras, H. Veltmeyer, and O. Nuñez, eds, *Globalización, Imperialismo y Clase Social*, Buenos Aires: Editorial Lúmen.

Stedile, Joao Pedro. 2000. Interview with James Petras, May 14.

Terceros, Walter and Jonny Zambrana Barrios. 2002. *Experiencias de los consejos de participación popular (CPPs).* Cochabamba: PROSANA, Unidad de fortalecimiento comunitario y transversales.

Toothaker, Christopher. 2007. "Chávez Cites Plan for 'Collective Property,'" Associated Press, Posted March 27. http://www.sun-sentinel.com/business/realestate/sfl-achavez27mar.

UNDP. 1996. "Good Governance and Sustainable Human Development," *Governance Policy Paper.* http://magnet.undp.org/policy.

UNICEF. 1989. *Participación de los sectores pobres en programas de desarrollo local.* Santiago, Chile: UNICEF.

Van Waeyenberge, Elisa. 2006. "From Washington to Post-Washington Consensus," in Jomo, K. S. and Ben Fine, eds., *The New Development Economics.* London: Zed Books.

Weisbrot, Mark. 2009. "The Chávez Administration at 10 Years: The Economy and Social Indicators," The Center for Economic and Policy Research (CEPR), Washington DC, February 5.

World Bank. 1994a. *The World Bank and Participation.* Washington DC: World Bank, Operations Policy Department.

———. 1994b. *Governance. The World Bank Experience.* Washington DC: World Bank.

5

Globalization and Marginalization of Labor: Focus on Sub-Saharan Africa

Johnson W. Makoba

At the 2002 meeting of world leaders in Monterey, Mexico, to discuss ways of fighting global poverty, most world leaders concluded that neoliberal globalization as a strategy for eradicating poverty in developing countries has proven disappointing. According to the perceptions of leaders of the most powerful advanced capitalist countries at the Monterey meeting, neoliberal globalization "...has done far less to raise the incomes of the world's poorest people than leaders had hoped..." (Weiner 2002, A6). Moreover, it is reported that "the vast majority of people living in Africa, Latin America, Central Asia and the Middle East are no better off today than they were in 1989 when the fall of the Berlin Wall allowed capitalism to spread worldwide at a rapid rate"[1] (Weiner 2002, A6).

Neoliberal globalization is a process that can be analyzed in terms of its several important aspects. First, it can be considered the latest, most advanced stage in the history of capitalism. Second, it can be seen as an outcome of the internationalization of capital via the activities of transnational corporations. Third, it entails production that can be transferred overseas by transnational corporations in order to drive out competitors. Although other powerful forces of globalization exist—for example, international financial institutions, such as the World Bank (WB) and the International Monetary Fund (IMF), and regional and global trade and economic groupings, such as the North American Free Trade Agreement (NAFTA) and the World Trade Organization (WTO)—the most dominant and visible agent of neoliberal globalization is the transnational corporation (TNC). The TNCs venture outside of their home territory to secure cheap labor, raw materials, new markets, and superprofits.

Neoliberal globalization is a powerful force responsible for the massive transformation of nation states, economies, international institutions, as well as creating a new world economic order. It is a process rife with internal contradictions. Thus, it creates winners and losers in the global balance of power. This occurs as some countries become increasingly integrated into the global system, while others become marginalized. The polarization among and between countries tends to parallel trends in global labor force bifurcation. The process of labor marginalization in developing countries began during the colonial era, but as we show below, it has intensified during the past three decades.

There is growing evidence that neoliberal globalization is contributing to increasing poverty and inequality around the world (Cox 1997; Kirkbride 2001; Nissanke and Thorbecke 2006, 1338–1342). In Africa, Asia, Latin America, and elsewhere, poverty and inequality are being intensified by the increasing marginalization and feminization of labor, especially in export-oriented manufacturing and agricultural sectors (McMichael 2008; Sassen 1998).

This chapter contends that neoliberal globalization has greatly contributed to increased inequality and poverty as well as to the marginalization and feminization of labor in developing countries, especially in Africa. The chapter is divided into four parts: the first part looks at the contribution of neoliberal globalization to inequality and poverty in Africa; the second part examines processes involved in the marginalization and feminization of labor in developing countries; the third part discusses the evidence concerning the marginalization and feminization of labor in export-oriented sectors of selected developing countries; and the fourth part briefly considers resistance of marginalized labor and other groups against neoliberal globalization and efforts to empower workers across the globe.

Globalization, Inequality, and Poverty in Africa

Two critical arguments revolve around the impact of neoliberal globalization in Africa, Latin America, and South Asia. Proponents of neoliberal globalization insist that it contributes to economic growth and the narrowing of the income gap between advanced capitalist and developing countries (Dollar 2001, 2–3; Friedman 2007, 251–252; Kayizzi-Mugerwa

2001, 6–7, 26; Stiglitz 2007, 295; Sacks 2007, 357).[2] In contrast, opponents maintain that it increases inequality and poverty in these countries. Not only are economic inequalities between advanced and less-developed capitalist countries growing rapidly, but they are growing or persisting "at very high levels, especially in...Africa" (Weissman 2003, 2). Proponents of neoliberal globalization, including the World Bank, acknowledge that "while the poverty rate in South Asia fell from 52 percent in 1981 to 31 percent in 2002, the poverty rate actually increased in sub-Saharan Africa from 42 percent to 44 percent and the number of people living in poverty nearly doubled" (World Bank 2007, 18–19). Moreover, the average daily incomes of the poor in sub-Saharan Africa (SSA) have not increased since the 1990s. It is reported that such incomes have remained "at a meager $0.62 a day—pointing to the severity and depth of poverty in this region" (World Bank 2007, 19). It is estimated that "no less than 340 million African people [out of 800 million] live on less than U.S. $1 per day" (Magbadelo 2005, 96).

Compared to other regions of the developing world, African countries have been hit the hardest because they are both the most economically dependent and marginalized within the global capitalist system. Neoliberal economic policies designed by the World Bank and the IMF to overcome Africa's economic crisis and integrate their economies into the global system have contributed to more poverty and inequality on the continent. The World Bank and IMF took advantage of Africa's economic crisis that began in earnest in the 1980s to pressure African and other states to abandon the statist and inward-looking strategies advocated by development theory and to embrace outward-oriented neoliberal economic policies stressed by these global institutions controlled by the advanced capitalist countries. The neoliberal policies imposed on African countries by the "Washington Consensus" have failed to deliver economic growth and the promised benefits. According to Arrighi et al. (2007), "such policies have been associated with a sharp deterioration of their economic performance [and] the median rate of...their per capita income falling from 2.5 percent in 1960–1979, to zero percent in 1980–1998" (Arrighi, Silver, and Brewer 2007, 330).

Overall, the neoliberal policies are perceived to have made the African economic crisis worse. A report issued by the Economic Commission for Africa (ECA) indicated that the nearly 30 countries implementing World Bank-IMF policies between 1990 and 1991 "faired worse economically than those without such programs" (Lancaster 1991, 94). According to Martin (2002, 38), "the growth rates of almost all [African] countries over the last three decades have been either negative or negligible." Evidence based on sectoral studies shows that neoliberal policies tend to "deindustrialize the existing manufacturing base in many African countries without encouraging any significant replacement" (Stein 1992, 83). Moreover, many African countries that implemented neoliberal policies also experienced "a widespread erosion of local peasant economies and social communities" (Bryceson 2002, 737). The deindustrialization process has forced African countries to rely on agricultural exports that are no longer subsidized by African governments, and yet they are vulnerable to demand and price fluctuations in the global economy.

Uganda, Ghana, and Tanzania are three African countries often cited as economic success stories that effectively implemented neoliberal policies. Since the 1980s these three countries have experienced high rates of economic growth (from increased aid), without corresponding income growth or poverty reduction. While the overall real GDP in Uganda grew by 6.3 percent between 2001 and 2003, growth in the agricultural sector was only 5.1 percent. The comparatively slower rate of agricultural growth in Uganda makes it difficult to achieve the country's objectives of poverty eradication, especially in rural areas where more than 80 percent of the people live.

Over the past ten years, Uganda has experienced gradual and sustained economic growth. However, the benefits of such growth have not been evenly distributed or shared (Ssewanyana et al. 2006). For example, between 1992 and 2000, there were more poor people concentrated in northern and eastern regions of the country than in the central and western regions that have generally benefitted from access to resources from the government, the private sector, and the nongovernmental organizations. Both the northern and eastern regions of Uganda, which have been politically and economically marginalized, saw the absolute number of the poor increase by 48 percent and 14 percent, respectively, over the same period. It is therefore clear that benefits of strong economic growth have not translated into reasonable improvements in the standard of living of majority of Ugandans. In fact, many Ugandans believe poverty is worsening in their communities.

Uganda's improved macroeconomic outlook is a result of massive aid inflows that have contributed to a huge external debt. It is reported that Uganda's "external debt grew from $1.3 billion in June 1987 to $2.9 billion at the end of June 1994" (Sharer, DeZoyza, and McDonald 1995, 6). In 2003, the external debt was estimated at more than $4 billion. The debt relief plan under the Heavily Indebted Poor Countries (HIPC) initiative saves the country $90 million annually. These savings are not adequate to meet the funding needs of education, health, and infrastructure. Despite the debt relief, Uganda's debt-to-exports level of 307 percent by the end of 2003 (instead of the recommended ratio of 150 percent or less) makes the initiative unsustainable (Kuteesa and Nabbumba 2004). As the external debt was increasing, Uganda's export earnings declined due to lower global market prices. For the year ending 2002–2003, export earnings fell by 28 percent, from a projected $1 billion to $726 million. The price of coffee, which brings in 60 percent of export earnings, has declined more than 70 percent since the late 1990s (Kuteesa and Nabbumba 2004). In addition to the decline in export earnings, direct foreign investments and domestic savings have not increased, as envisaged under the neoliberal policies. More importantly, since capital outflows continue to exceed capital inflows, sustainable economic growth has not been achieved or continues to be elusive more than two decades after the implementation of neoliberal policies in Uganda.

In Ghana, the poor economic performance of its agricultural sector from the 1980s to the present has been blamed on neoliberal economic and trade policies. In particular, ending state subsidies, abolishing the state licensing system for imports (including agricultural food imports), and the reduction of tariffs have meant that Ghana's small farmers are exposed to global competition from cheap imports. Small or medium size commercial farms in Ghana that were engaged in production as viable enterprises were undermined as they were exposed to competition from agricultural imports from advanced capitalist countries (such as the United States and the European Union) that normally subsidize agricultural production (Khor 2006, 1–2, 20–21). Despite the neoliberal rhetoric of reform, the volume of resources going to the agricultural sector declined sharply throughout the 1990s. For example, "the share of agriculture in public expenditure... declined from 27.3 percent in 1990 to 10.8 percent in 1998" and the "overall performance in agriculture in general (and food agriculture in particular) has lagged behind all other sectors, with growth averaging 2.5 percent during the 1990s" (Khor 2006, 11). And prices of agricultural commodities for export have declined, reducing foreign exchange earnings. The price of cocoa, Ghana's main export, declined by 14 percent between 2002 and 2005 (Economic Commission for Africa 2007, 23–24). Continued poor performance of the agricultural sector and increased food imports from advanced capitalist countries have displaced locally produced agricultural products (such as rice, toma-toes, and poultry). This has greatly contributed to rural poverty and great rural-urban inequalities in Ghana (Afrobarometer 2008).

Despite strong growth for nearly fifteen years, both poverty and income inequality between rural and urban areas have been increasing. A recent survey shows that a majority of Ghanians report short-ages of cash and hold the position that neoliberal policies pursued by the government have hurt most people (Afrobarometer, 2008). The income inequality between rural and urban areas and across Ghana's administrative regions has increased greatly, as shown by the Gini Coefficient, which increased from 0.353 to 0.394 over the past fifteen years (Ghana CEM, 2007). Beyond the increase in pov-erty and inequality, Ghana's external debt has increased to almost $7 billion. And 20 percent of Ghana's export earnings are used to service the national debt. Ghana has not yet qualified for the debt relief plan under the HIPC initiative, as is the case with Uganda and Tanzania.

The sharp decline of production in the agricultural and indus-trial sectors in Tanzania in the 1970s due to both adverse internal and external factors led to a serious economic crisis in the 1980s. President Julius Nyerere's government responded to the economic crisis between 1982 and 1984 "by emphasizing greater economic incentives for agricultural producers, a reduction in the government's large budget deficits and a need for greater prudence in the manage-ment of the country's money supply" (Makoba 1998, 212). However, because of the severity of the economic crisis, these measures were not enough to turn the economy around. Nyerere, while still president of Tanzania, fiercely resisted the tough neoliberal economic policies of the World Bank and the IMF.

But when Nyerere stepped down as president in 1985, the new president, Ali H. Mwinyi (his successor), and the IMF accelerated and broadened neoliberal economic reforms in Tanzania. Observers believe that when the president resigned in 1985, the most ardent opponent of the World Bank and the IMF had disappeared from the political pulpit. His departure inevitably gave the pro-IMF fac-tion among the technocrats and the Chama Cha Mapunduzi (the ruling CCM or Revolutionary Party) momentum to conclude and sign a Structural Adjustment Agreement with the IMF in 1986. Both the IMF and the World Bank demanded that Tanzania imple-ment a Structural Adjustment Program (SAP) designed to reduce the role of the state via trade liberalization and privatization of major state-owned enterprises (or parastatal organizations). The neoliberal policies imposed by the IMF "forced the Mwinyi government to shift farther away from Tanzania's objectives of socialism and self-reliance" (Makoba 1998, 214) that were vigorously pursued throughout the late 1960s and early 1970s. They also increased inequalities between the elite and ordinary people as the Leadership Code intended to prevent political leaders from owning businesses or receiving more than one income was abandoned by the Mwinyi government. According to the most recent IMF report, economic growth performance has been "on track although it slackened slightly to 6.2 percent in 2006, compared to 6.8 percent attained in 2005" (International Monetary Fund 2008a, 7). Increasing empirical evidence seems to suggest that even high rates of economic growth do not translate into poverty reduction or lesser inequality. "In Tanzania, like in many other developing countries, there is growing concern that improved economic growth performance is failing to lead to improved well-being for the poorer citizens" (Mbele 2007, 10). In other words, improved economic growth in Tanzania has not led to improved well-being as anticipated under neoliberal reform policies adopted more than two decades ago. This is, in part, due to the slower economic growth in the agricultural sector where the majority work or earn their livelihoods. For example, in 2004, sectors of the Tanzanian economy that grew fastest and fueled the real GDP growth averaging 6.7 percent in 2005 were mining (15.6 percent), construction (12.0 percent), and tourism (8.0 percent), while agriculture grew at the slower rate of 5.9 percent (Tanzania Social and Economic Trust 2006, 2). This means that economic growth in the agricultural sector, which accounts for 50 percent of GDP, 85 percent of exports, and employs 80 percent of the Tanzanian work force, has been growing at a slower pace compared to other sectors of the economy. It is therefore self-evident why the majority of Tanzanians are not very much impacted by neoliberal economic policies pursued at the national level.

The fact that national macroeconomic gains have not translated into poverty reduction shows that the growth process or neoliberal policies in general are not pro-poor or antipoverty. A recent IMF Tanzania Country Report seems to acknowledge that there is no visible impact on poverty reduction or improved well-being/quality of life of the population despite increased growth in GDP (International Monetary Fund 2008b, 89–90). Tanzania, like other sub-Saharan African countries, "is caught up in a process by which previous structural adjustment conditions have been replaced by the Poverty Reduction Strategy Paper (PRSP)" (Ellis and Mdoe 2003, 1367).

Increasingly, poverty reduction strategies are being implemented in Tanzania and elsewhere in sub-Saharan Africa to "mitigate" the negative consequences of the structural adjustment programs. As a result, "rural development strategies have shifted focus in Tanzania today from growth maximization and agricultural intensification to poverty reduction" (Erdal 2005, 2). Currently, Poverty Reduction Strategy Paper (PRSP) serves as the leading framework for poverty reduction and rural development in Tanzania and other sub-Saharan African countries.

Rural-urban poverty and inequalities in Tanzania are reported to have increased between 1991 and 2001. According to Ellis and Mdoe (2003, 1369):

> Poverty in Tanzania is much worse in rural than in urban areas, with an estimated 39% of rural citizens being poor compared to 24% of urban citizens in 2000–01. The lowest incidence of poverty is in Dar es Salaam where the ratio falls to 18%. Income inequality in Tanzania appears to be increasing, the Gini Coefficient for the country as a whole rising from 0.34 to 0.37 between 1991–92 and 2000–01, and for Dar es Salaam from 0.30 to 0.36.[3]

It is argued that the ending of subsidies and other support provided to rural areas by the government due to neoliberal policies has had a negative impact on rural productivity and livelihoods. Views expressed by Tanzanians in recent surveys paint a much gloomier picture of deepening poverty and growing social inequality (Cooksey 2004, 2). For example, it is reported in a survey covering 3,000 households in seven regions of Tanzania that "Only ten percent of the respondents thought that all Tanzanians have benefited more or less equally from the government's economic reforms; 90 percent thought only a minority have benefited, while for most people things are as bad as or worse than before" (Cooksey 2004, 2).

In a 2001 survey conducted by the University of Michigan-based Afrobarometer, "60 percent of respondents agreed with the statement: 'the government's economic policies have hurt most people and only benefited a few.' Large majorities thought public policy performance was bad in ensuring food security, job creation, reducing poverty and the rich-poor gap" (Cooksey 2004, 2). There are several explanations that account for the poor economic performance of the rural agricultural sector and persistent rural poverty. Some of the most important reasons are discussed here. First, "a growing number of Tanzanian officials and academics argue that "structural adjustment dealt a body blow to farmers by liberalizing import and export markets, with the result that fertilizer has become prohibitively expensive for farmers and farm-gate prices are low and volatile" (Cooksey 2004, 5). Various research findings have shown that farmers blame both liberalization policies and corrupt private buyers for all their problems. Second, the poor performance of reorganized marketing boards combined with inadequate government capacity to supervise them or provide adequate inputs and extension services to the farmers have contributed to poor performance of the agricultural sector. Tanzanian farmers are reported to have said that "the major barrier to the low usage of inputs was lack of easy availability and many blamed this to the disappearing of cooperatives [replaced by new marketing boards] that earlier had been the traditional source of agricultural inputs" (Erdal 2005, 29). A decline in both the quality and access to agricultural extension services denied farmers the critical know-how necessary for modern farming techniques and services necessary for improved productivity. The third reason for the poor performance of the agricultural sector is attributed to lack of credit. According to Erdal (2005, 30), "credit for agricultural purposes has declined sharply, nearly collapsed, since 1996. The share of loans for agricultural credit fell from 19.7% in 1995 to 0.8% in 1999." The dismantling of cooperatives and rural development banks demanded by the IMF compounded the problem of rural agricultural credit even further. Finally, liberalized markets benefited rich farmers and new cooperatives that are controlled by the rich. Besides being politically well-connected and "setting prices" for rural produce, the rich farmers have greater economic advantages (Erdal 2005, 31). As a result, neoliberal policies intended to create an efficient rural agricultural system based on market forces have yielded greater profit for rich farmers while undermining the agriculture sector in general, leading to increased rural poverty and inequality.

Since the 1970s, Tanzania has borrowed extensively from both bilateral and multilateral sources. In particular, "extensive bilateral lending in the 1970s and multilateral lending tied to the IMF and the World Bank conditions in the 1980s and 1990s has not brought about human and social development." Critics contend that "extensive borrowing has contributed to building up a huge unsustainable debt." Tanzania's debt burden has increased dramatically since the 1970s. And with the increase in debt, debt servicing has grown over the past three decades. It is reported that "the external debt stock increased from U.S. $197 million in 1970 to U.S. $5.2 billion in 1980 and U.S. $7.8 billion in 2004. Debt service grew from only U.S. $3 million in 1970 to U.S. $114 million in 2004." The servicing of the debt absorbs about 40 percent of total annual government expenditures. While development expenditures declined in the 1980s, "in the 1990s spending on debt was higher than spending on health and education combined."

Any modest gains in indicators for human and social development in Tanzania have largely come about as a result of increasing aid levels and declining debt service. The declining debt service is due to the debt relief plan under the Heavily Indebted Poor Countries (HIPC) initiative, which requires a "home grown" national poverty reduction strategy as a condition for implementation. As a result of the debt relief plan, the Paris Club between 1990 and 2000 forgave $1.6 billion of its debt and rescheduled another $1.3 billion of the same. However, these debt deals have not freed enough resources for human and social development. In addition to borrowing new loans to finance up to 40 percent of its national budget annually, Tanzania still needs to pay millions of dollars a year to foreign non-Paris Club creditors. Full cancellation of its external debt would be the right solution to its debt problem.

The negative impact of neoliberal policies on the economic performance of Uganda, Ghana, and Tanzania can easily be extended to all African countries. It is reported that "after nearly two decades of pursuing IMF/World Bank structural adjustment programs, almost half the population of the African continent lives in grinding poverty" (Worell 2001, 45). Increasingly, Africa's debt problem is constraining economic growth and improvements in people's quality of life. Sub-Saharan African countries owe more than $200 billion to foreign creditors, mostly to the IMF and the World Bank. These countries must pay interest on the loans and pay back the principal, whether the loans are invested productively or not. Sub-Saharan Africa pays more than $10 billion annually in debt payments. This means that funds that could otherwise be used for education, health care, clean water, or infrastructural development are siphoned out of Africa. It is reported that "for the poorest indebted countries, a group of 42 heavily indebted poor countries (HIPC), mostly concentrated in Africa, which owe most of their debt to official creditors, the IMF and World Bank have fashioned a modest relief program" (Weissmann 2003, 3). Under the debt relief, more than half of the HIPC countries have received some relief. However, given Africa's economic crisis and lack of financial resources, what is urgently needed is "debt cancellation" rather than debt relief. Global grassroots advocacy groups such as "Jubilee Plus" and "Drop the Debt" are spearheading efforts to insure that Africa's huge debt is cancelled. Whether these efforts will succeed or not is too early to tell.

Neoliberal policies in both Africa and Latin America have contributed to more inequality and poverty. Since the 1980s, poverty rates in Africa have worsened while those in Latin America have stagnated. It is reported that "while Africans accounted for only 16 percent of the world's deep poverty in 1980, they made up two-thirds of the poor in 1998" (Lerman 2002, 3). In contrast, poverty in Latin America was not as severe as the African poverty over the same period. In reality, both Africa and Latin America neither gained nor lowered their overall poverty rates in the past two decades. "Today, 80 percent of the world's population lives in countries that generate 20 percent of the world's total income" (Goulet 2002, 8). The world economy is not poor, it is simply an "extremely unjust and unequal economy..." (Goulet 2002, 9).

The Processes of Marginalization and Feminization of Labor in Developing Countries

Marginalized labor tends to be low-skilled and low-paying, with no job security or benefits. The jobs performed by predominantly single young women are carried out under conditions closer to those of servitude than meaningful or legitimate employment. Cheap labor–producing countries are part of a new international division of labor dictated by the global economy.

Neoliberal globalization is inevitably creating a polarized global labor force. Such a labor force includes a core of relatively stable, better-paid workers in advanced capitalist countries and poorly paid, marginalized workers, mostly women, in the developing countries. The polarized or bifurcated labor force is both an outcome and a relational process, where core and marginalized labor forces are interdependent. As firms in advanced capitalist countries restructure and embrace the so-called "lean mean production" strategy, they tend to eliminate less-skilled jobs and fill tasks associated with them through subcontracting or outsourcing to nonunionized labor or relocating production to other countries where cheap, casual labor, employing mostly women, is abundant.

McMichael points out that "the U.S. automobile sector outsourced so much of its component production beginning in the late 1970s that the percentage of its workforce belonging to unions fell from two-thirds to one-quarter by the mid-1990s" (2000, 191). The impact of outsourcing on the U.S. auto industry had a very devastating impact on the labor force. First, the outsourcing polarized or bifurcated the labor force into better-paid and poorly paid workers. Second, the use of nonunion workers as a result of outsourcing components of production eroded wages of unionized workers.

Corporations in the United States and other advanced capitalist countries were able to restructure their work force and relocate some semi-skilled and unskilled jobs abroad. However, countries in Africa, Asia, Latin America, and the Caribbean, lacking the technological capacity to create high-skilled jobs, saw instead the expansion in low-tech, semi-skilled and unskilled, labor-intensive jobs, especially in the garment and electronics assembly industries. The majority of workers in these low-tech, labor-intensive sectors are single women. It is reported that between the mid-1970s and mid-1990s "the garment production sector alone spawned 1.2 million jobs in Bangladesh, 80 percent of which [were] held by women":

> women tend to enter the new industrial jobs, such as in *maquiladoras* in Mexico, where they constitute two-thirds of the labor force; and in global factories in East Asia, where they average 42 percent of the work force. In Latin America in particular...this trend of *feminization of global labor* parallels the downgrading of male employment as many formal jobs are restructured and turned into casual employment. (McMichael 2000, 191)

The casualization of employment relations in the global economy represents the expansion of what are typically considered casual or unsheltered jobs and a dramatic growth in part-time jobs, including high-paying professional jobs. This process weakens and even eliminates the claims by workers on companies that employ them, and thus weakens the overall position of labor in the economy. It is estimated that "over sixty percent of all part-time workers in the U. S. labor force are in labor-intensive services, which is also the sector that is expected to add the largest share of new jobs over the next decade" (Sassen 1998, 146). The phenomenon of casual labor, though caused by different factors, occurs throughout the global economy—in advanced as well as less-developed countries. In Africa, Latin America, and elsewhere, casualization of labor is caused by the process of "depeasantization." This process entails peasants being expelled from the land to make room for the expansion of export-oriented agriculture. The landless peasants inevitably migrate to urban centers/cities in search of paid or wage labor. This process that has been going on for centuries in advanced capitalist countries has recently accelerated in the developing countries, as more areas of arable/cultivatable land and forests are used for export-oriented agriculture, which is an integral part of the global market. The pools of labor resulting from depeasantization create the low-cost and casual labor forces sought by transnational corporations interested in outsourcing operations worldwide. In advanced capitalist countries such as the United States, casualization of labor is created by unstable employment conditions resulting from competitive labor markets (McMichael 2000, 192). As transnational corporations using the "lean, mean strategy" retool, restructure, and relocate overseas or go out of business, they shed labor without alternative employment opportunities.

Neoliberal globalization links depeasantization in developing countries with casual labor in advanced capitalist countries in a unique and interesting manner:

> As peasants lose markets to cheaper imported foods or surrender their land to larger commercial agro-export operations, they flood the towns and cities looking for work. When barriers to trade and investments fall [as in the case of NAFTA], the cheaper labor these peasants can provide attract foreign investment as firms scour the world, or the region, to reduce production costs [and reap super profits]. (McMichael 2000, 192)

Thus, as Kirkbride points out, it is ironic that "while organizations were slimming down to their core activities [due to restructuring], globalization pressures were encouraging them to expand their geographical reach, so that the new organizations became more global and more focused" (Kirkbride 2001, 76–77).

Transnational corporations in textiles and electronics consider the greatest source of cheap labor to exist abroad. Hence, corporations from advanced capitalist countries relocate in countries around the world to reduce their labor costs and reap superprofits. Governments of developing countries interested in promoting economic growth and generating employment in the export-oriented manufacturing sector tend to create special economic zones, lower taxes, and prevent unionization in a bid to control labor costs and social unrest. In some cases, these countries compete among themselves to provide the lowest wages in order to attract transnational corporations. For example, in order "to facilitate the establishment of assembly plants, governments in Indonesia, Malaysia, Guatemala, and Mexico...created in their countries *free trade zones*, areas in which large corporations were permitted to deliver goods to be assembled—cut cloth for wearing apparel, electronic components, and so on—and for which they were not required to pay tariffs...." (Robbins 2002, 46). Large developing countries in Africa, such as Nigeria, Egypt, or South Africa, have developed "pools of cheap labor" for transnational corporations. In exchange, the transnationals have agreed to hire local workers at low wages. The home countries of the transnationals, such as the United States, pass legislation that permits the corporations to transfer the assembled goods to the United States, paying taxes only on the labor cost of each product rather than on the total value of the product.

Transnational corporations seek female workers for their off-shore operations for three major reasons. First, young single women between the ages of sixteen and twenty-four are seen as willing to accept a wage level less than that needed for subsistence (or a "living wage"). Assembly plants in developing countries prefer single women because managers believe that older, married women have too many obligations, are often unwilling to work night shifts, and may not accept low wages. Women and children, who are the most marginalized, constitute a socially vulnerable work force to secure low-paid labor.

The second reason women are preferred for employment is that they are assumed by management to be easy to manage, control or discipline, and are less likely to join unions. Because women occupy the lowest positions in assembly plants, they can easily be hired and fired, depending on the overseas demand for textiles, shoes, plastics, electrical appliances, and, increasingly, electronics. Factory managers, with the support of male relatives of female workers, use traditional forms of social and cultural control to enforce discipline among female employees.

The third reason for employing women is the absurd claim by some managers that assembly plant work is designed with women in mind. Some managers go to the extent of claiming that women are better able to concentrate on routine, boring work than men! Other managers insist that young girls have better eyesight and are better at adjusting to assembly work.

The process of targeting women for low-wage jobs does affect the meaning that societies give to specific tasks. Very often, the division between skilled and unskilled jobs or tasks may not be based on the nature of work, but on who is doing the work. Such a distinction parallels the racial segmentation of work in the United States between the Irish and Blacks during the nineteenth century.[4]

Historically, free labor has been created by forcing people from the land or destroying their subsistence means of living. In different countries this was accomplished in different ways. However, "the overall result has been the creation of [marginalized] populations whose sole means of support is in the sale of labor" (Robbins 2002, 57). As we show in the next section, "industries and enterprises that depend on a cheap labor supply are able to take advantage of social divisions and discrimination that generally follow lines of gender, race, age, and country of origin to minimize labor costs and control the labor force" (Robbins 2002, 57). It is argued that capitalists may not necessarily create discrimination based on social divisions in societies, but capitalism certainly reinforces and benefits from such social divisions.

The Prevalence of Marginalization and Feminization of Labor in the Export Processing Sector

There is growing evidence of marginalization and feminization of labor in the export processing sectors of developing countries. Both governments and transnational corporations have been responsible for export expansion in these countries. Export growth in both manufacturing and agricultural products was part of the government strategy of export-led industrialization in these countries. Earlier exponents of this strategy, the East and Southeastern Asian NICS (i.e., South Korea, Taiwan, Singapore, and, more recently, Malaysia), managed to combine neoliberal policies of foreign direct investment (FDI) with political authoritarianism, low wages, and repression of labor to achieve rapid economic growth. These countries were able to attract foreign capital because of being on the frontlines of the East-West cold war rivalry, and they also promised political stability to investors. Their rapid economic success served as a model of economic development for other countries that followed in their footsteps. As a result, several countries especially in Asia and Latin America, including China, Indonesia, Malaysia, and Mexico, have created *free trade zones* to facilitate the establishment of export-manufacturing plants and offered transnational corporations extensive benefits such as tax breaks, cheap labor, and infrastructural support.

Export expansion was also part of a global strategy used by transnational corporations. This occurred as the transnationals discovered that "they could easily tap into pools of cheap labor by relocating their manufacturing processes...to countries on the periphery of the World System whose governments were committed to economic development through industrialization" (Robbins 2002, 46). Through this strategy, the transnationals became engaged in building "a new economic order" in manufacturing, agriculture, and services. In the process, global sourcing merged with the export-oriented strategy.

The policies pursued by the World Bank and the International Monetary Fund (IMF) facilitated the expansion of the export sector in developing countries. The actions of governments and transnationals supported by World Bank/IMF policies led to a dramatic growth in assembly plants in some of these countries. From the 1970s to the present, there has been a rapid increase in export-oriented assembly plants stretching from Asia's export processing zones to Mexico's *maquiladoras*. In Mexico, "the number of *maquiladoras* grew from virtually none in the 1960s to 1,279 employing 329,413 workers in 1985, to over 4,000 employing over one million workers in 1999..." (Robbins 2002, 47). In El Salvador, there were about 70,000 garment workers in 2001, 80 per cent of whom were women (Kaufman and Gonzalez 2001, A10). By 2004, the number of women workers in the Salvadorian garment industry had increased to approximately, 85,000 (Eriksson 2004, 9).

China has become a leading center for low-wage production in the global economy. The Chinese Communist Party under Deng Xiao Ping anticipated this development by establishing special economic zones in coastal regions in the 1980s to attract foreign direct investment. In the 1990s, as the East Asian NICs emerged as middle-income countries with high-skilled labor forces, China became the prime location for foreign investors from the United States, European Union, Japan, as well as Korean and Taiwanese investors who were experiencing rising labor costs at home. Today, China produces about half of the world's shoes and an array of electronic items, toys, and garments for the global economy. Local farmers in special economic zones in coastal regions of China now live off the rents from the factories. At the same time, tens of thousands of migrants from China's poorer hinterlands swell the low-wage work force. Neoliberal policies to privatize state-owned enterprises and China's recent entry into the World Trade Organization (WTO) are contributing to high rates of unemployment in the cities and the countryside and appear to threaten social and political stability.

Employment in the export-manufacturing sector acts as a double-edged sword. On one hand, the sector provides badly needed employment opportunities, especially to women, but working conditions are poor, pay is low, and governments attempt to control and repress labor, on the other. It is reported that "in some assembly factories in El Salvador, where women earn $4.51 for the day, or 56 cents an hour, union organizers are often summarily dismissed, bathrooms are locked and can be used only with permission..." (Robbins 2002, 47). In Guatemala, workers are required to work overtime at a moment's notice and, if they refuse, face dismissal. In American Samoa, United States labor laws and the federal minimum wage do not apply. Immigrant workers, especially Vietnamese, work under conditions of servitude. In Bangladesh, most garment factories tend to be "bleak, stuffy places with cramped aisles that dead-end into haphazard knolls of fabric" (Bearak 2001, A12). As one critic has observed about Bangladesh:

> What [it] has to offer the global economy is some of the world's cheapest labor—and what this impoverished nation has received in turn, is the economic boost of a $4.3 billion apparel industry, the fuller pockets that come with 1.5 million jobs and the horrors that arise from 3,300 inadequately regulated garment factories, some of which are among the worst sweatshops ever to taunt the human conscience. (Bearak 2001, A1)

The United States is Bangladesh's number one apparel customer. It is reported that nearly 50 percent of Bangladesh's $1.6 billion in garment exports end up in the United States annually. Efforts to improve working conditions in assembly plants have not been successful. This is because governments in such countries tend to collaborate with transnational corporations and their local management to exploit workers and suppress efforts aimed at unionization. It is reported, for example, that there is "systematic violence against union organizers in Mexico, El Salvador and Guatemala" (Robbins 2002, 47), as well as elsewhere.

Export-oriented agribusiness is modeled after export-led manufacturing. The food trade, which includes processed foods like meat, flour products, and fresh and processed fruits and vegetables, constitutes one of the fastest growth industries. Global food corporations increasingly organize producers on plantations and farms to deliver products for global markets. The global fruit and vegetable industry depends on flexible and cheap contract-labor arrangements. Like the export-manufacturing plants, agro-export production relies heavily on female labor. The Sayula Plant in Mexico employs "hundreds of young women whom the company moved by seasons from one site to another as a kind of mobile *'maquiladora...'*" (McMichael 2000, 201).

Global food corporations, like transnational manufacturing corporations, use global sourcing strategies. The production of the "world steer" [beef used in hamburgers] resembles the production of the "world car." It is produced in several locations with global inputs and sold in standardized packaging worldwide. It is reported that "from conception to slaughter, the production of the steer is geared entirely to the demands of the global market" (McMichael 2000, 101). International financial agencies such as the World Bank, the United States Agency for International Development (USAID), and the Inter-American Development Bank (IADB) play an important role in promoting livestock production by financing and requiring the establishment of cattle infrastructure in several Central American countries. The world steer production in Central America has:

> redistributed cattle holdings and open-range woodland from peasants to the ranchers supplying the export packers. More than half the rural population of Central America (35 million) is now landless or unable to survive as a peasantry. World steer production not only reinforces inequality in the production regions but also threatens craftwork and food security. (McMichael 2000, 102)

The spread of the world steer industry supplies the global market with beef, but at the same time undermines local agro-ecologies and displaces stable peasant communities, turning peasants into wage laborers. No doubt, world steer production reinforces inequality and poverty in producing regions of most Central American countries involved in beef production for the global market (popularly called "the hamburger connection"). It is estimated that one-tenth of American burgers use imported beef, much of it produced under contract by Central American meat packing plants for transnational food companies.

Transnational corporations, whether involved in export-manufacturing or agro-export production, often employ women. This is because women are perceived by management to be inexpensive, reliable, and easy to control. At the same time, governments of developing countries that seek to attract investments from transnational corporations in the export processing zones tend to offer numerous incentives. In many cases, such governments try to outcompete each

other by offering the lowest wages and the best package of concessions to transnational corporations. In this way, they collude with
transnational corporations in the marginalization and feminization
of labor across the globe.

Resistance Against Neoliberal Globalization

Neoliberal globalization has created serious consequences for states,
groups, workers, and individuals in many countries around the world.
However, the focus here is on the impact of neoliberal globalization
on marginalized labor and the response of labor and other social
groups to this phenomenon. Resistance or social protest against neoliberal globalization takes many forms, including unionization and
social movement mobilization.

Labor union decline in advanced capitalist countries such as the
United States is a direct result of outsourcing of service operations and
the lean, mean strategy of corporate restructuring and downsizing in
pursuit of efficiency and superprofits within the global economy. At
the global level, labor's response to corporate restructuring has been
to "forge new forms of organization, such as the new labor internationalism that has emerged to present a solid front to footloose firms
that would divide national labor forces, and to states that enter free
trade agreements that would undermine labor benefits" (McMichael
2000, 196). In other words, as neoliberal globalization gives corporations the freedom to move around the world seeking cheap labor and
lax regulations, it enables organized labor the opportunity to fight for
improved working conditions.

This new labor internationalism is praised for contributing effectively to the political debate surrounding NAFTA. It is reported that
organized labor in the United States led by the rank and file distanced
itself from the U.S. national policy of free trade, arguing that NAFTA
was not in the best interests of U.S. workers. Thus, organized labor
"joined a substantial national political coalition of consumers, environmentalists, and others in opposing the implementation of NAFTA"
(McMichael 2000, 196). Moreover, progressive unions on both sides
of the U.S.-Mexican border have come together in cross-border organizing and unionization efforts to protect workers' interests. Because
of these actions, "the stranglehold of the Mexican government on
union organization has begun to fray, [as] evidenced by the formation
of an independent union, the Authentic Labor Front, which formed
an alliance with the U.S. United Electrical Workers, Teamsters,

Steel Workers, and four other U.S. and Canadian unions in 1992" (McMichael 2000, 196). This development parallels labor movements in other countries around the world where independent unions are emerging in the new environment of global integration. Organizations facilitating the emergence of autonomous unions in these countries include the revamped Transnationals Information Exchange (TIE). TIE is "dedicated to forging networks of labor organizations across the world, which it pioneered in connection with the 'global factory' associated with the production of the 'world car'" (McMichael 2000, 196). In one of its innovative projects, called the Cocoa-Chocolate Network, TIE used a novel strategy of the "production chain" that "linked European industrial workers with Asian and Latin American plantation workers and peasants, extending the production chain back from the chocolate factories to the cocoa bean fields" (McMichael 2000, 197). TIE, modeled as an international social movement, has created a new form of labor internationalism "connecting casual labor across national boundaries, organizing regionalized networks of labor, and addressing issues of racism and immigrant workers" (McMichael 2000, 197). Thus, through its global network, TIE has been effective in empowering otherwise marginalized labor in Asia, Latin America, and elsewhere.

Resistance and social movements are not restricted to global labor or the global economy. New social movements are bringing pressure to bear in global civil society as well. The globalization of civil society entails resistance from disadvantaged groups in a changing division of labor. Such resistance may be organized by the groups themselves or championed by nongovernmental organizations (NGOs) that are dedicated to democratization and protection of the human rights of marginalized and disadvantaged groups such as women, children, immigrant and casual workers. NGOs, especially those engaged in economic empowerment, democratization, and human rights advocacy, are vital in transforming masses from being faceless to being recognized as powerful actors with interests and needs to be met.

As a result of the declining power of organized labor and revolutionary groups due to global restructuring and the end of the cold war, marginalized labor and other disadvantaged groups must redefine their role in the emerging order. In other words, the powerless must devise alternative strategies of social struggle. These groups should strive to use their popular participation in grassroots organizations to assert control over authoritarian governments locally and the seemingly remote, but powerful forces of globalization that are operating on a world scale. Under the pressure of neoliberal globalization, state legitimation crisis is increasing throughout the world, due in part to state indifference or incapacity to provide services or resolve the breakdown of social institutions. Such weakened states increasingly face disillusioned citizens, repressed and exploited workers, and neglected local communities that demand change. Movements for democracy or oppositional groups have emerged putting more pressure on the overextended, debt-ridden, authoritarian and corrupt states. New social movements, such as women's groups, environmental, human rights organizations, or the World Social Forum, are sometimes linking up with organized and unorganized labor to confront threats posed by WTO, the World Bank, IMF and the G-3 (i.e., United States, Europe, and Japan). Such movements seek either "a different kind of globalization" (Goulet 2002, 2) or "to bring better-paying jobs to neighborhoods and to ensure that people have access to healthcare, education, and clean communities in the United States and abroad" (Hytrek and Zentgrab 2008, 164). As a result of the pressure exerted by the antiglobalization movements, Horst Kohler, the former Managing Director of the IMF, acknowledged that the fund had made mistakes in the past and called on the demonstrators to participate in a constructive dialogue. In a similar way, James D. Wolfensohn, the former World Bank President, is quoted as having said that 'the bank had become more open and more responsive to its critics in the past five years, including meeting with some of the critics. It is too premature to celebrate the formation of a unified global social movement designed to counteract the deleterious effects of neoliberal globalization. This is in part because such a global social movement is yet to coalesce.

Finally, other forms of resistance to neoliberal globalization, such as consumer advocacy, community-supported agriculture, organic food systems, and fair trade groups, have increased pressure on both transnational corporations and governments around the world. One of the most effective advocacy groups has been the United Students against Sweatshops (SWAS) formed in 1998 and supported by about 160 U.S. colleges and universities. SWAS is against the link between U.S. universities and offshore sweatshops in producing logo-emblazoned clothing using low-wage labor under poor working conditions. It is reported that, as a result of SWAS' efforts, Nike "has raised wages of its workers in Indonesia, even though the [Asian] financial crisis undermined their real purchasing power" (McMichael 2000, 274).

Concluding Remarks

Proponents of neoliberal globalization see it as an attempt to construct a new economic order that would serve as the blueprint for global development in the twenty-first century. However, critics blame neoliberal globalization for increasing inequality and poverty. Neoliberal globalization is creating fundamental changes in the global order. Increasing evidence seems to point to widening inequalities within and between countries. Also, some groups or classes within given countries appear to benefit while others are marginalized. This is because the effects of neoliberal globalization are inherently uneven at national and global levels.

Neoliberal globalization is inevitably creating a polarized global labor force. Such a labor force includes a core of relatively stable, better-paid workers in advanced capitalist countries or regional centers of transnational operations and poorly paid marginalized workers in developing countries. Both export-oriented manufacturing and agro-production sectors in these countries employ low-cost wage labor, especially of single women. The increased incorporation of women in export-manufacturing and agro-export production has intensified the feminization of marginalized labor. Female labor is popular in both sectors because it is considered reliable, cheap, and easy to control. Depeasantization—expulsion of peasants from the countryside to urban areas/cities—has increased the ranks of the urban unemployed who are desperately in search of work. Global food corporations, like manufacturing corporations, benefit from low-cost labor provided by desperate landless peasants and socially vulnerable single women. It has been argued that the greatest force in the globalization of labor markets is the unlimited access to cheap labor by transnational corporations operating abroad, especially in such large countries as China, India, Brazil, Nigeria, Egypt, Indonesia, and Mexico. Transnational corporations relocate operations to such countries in order to exploit low-wage labor and thereby increase profits. For their part, governments in these countries compete in offering low wages and other packages to attract foreign direct investment in specially designated geographic areas known as export processing zones.

Resistance or social protest against neoliberal globalization is on the rise. This includes unionization or participation in new social movements that revolve around the politics of culture and identity. The intensification of labor and nonlabor countermovements against neoliberal globalization reveals that this phenomenon is a social process with inherent contradictions that require resolution. As Kirkbride points out: "One challenge for the twenty-first century is whether globalization will be as effective in providing opportunities for suppliers of labor as it was in the twentieth century in providing opportunities for the suppliers of capital" (Kirkbride 2001, 91). Historically, transnational corporations or owners of capital have never provided good working conditions and benefits to labor; rather, they have exploited and alienated labor. Thus, labor will have to be united and organized in order to fight for improved wages and overall working conditions in the new global economic order.

Notes

1. Jan Pakulski (2004:181) adds that "globalization...seems to widen socioeconomic gaps within the countries that follow a rapid deregulation and capital redistribution."
2. Joseph E. Stiglitz, argues that "the problem is not with globalization, but how it is managed" (2007, 295). And Jeffrey D. Sacks claims that "Africa's problems are not caused by exploitation by global investors, but rather by its economic isolation, its status as a continent largely bypassed by forces of globalization" (2007, 357).
3. Japan Bank for International Cooperation in its 2006 Report concludes that "in terms of the size of the population of the poor, the population below the food poverty line and the CBS poverty line in fact increased" (2006, 1).
4. According to Robbins, "The key to the distinction between White and Black [in the United States] became work; White meant doing 'white man's work,' while Black meant doing 'Black man's work.' The distinction was arbitrary because many jobs that became White man's work when reserved for Irish had been performed by Blacks earlier. This distinction resulted, then, in a situation in which to be 'White' the Irish had to work in the jobs from which Blacks were excluded..." (2002, 41).

References

Afrobarometer Briefing Paper No. 50. June 2008. "Economic Conditions in Ghana in 2008." www.afrobarometer.org.

Arrighi, Giovanni, Beverly J. Silver and Benjamin D. Brewer. 2007. "Industrial Convergence, Globalization, and the Persistence of the North-South Divide," in Roberts, J. T. and A. B. Hite, eds., *The Globalization and Development Reader: Perspectives on Development and Global Change*. Malden, MA: Blackwell.

Bearak, Barry. 2001. "Lives Held Cheap in Bangladesh Sweatshops," *The New York Times*, April 15.

Beynon, John and O. Dunkerley (eds.). 2000. *Globalization: The Reader*. New York: Routledge.

Bryceson, Deborah F. 2002. "The Scramble in Africa: Reorienting Rural Livelihoods." *World Development* 30, 5.

Bumiller, Elizabeth. 2002. "White House Letter; Diplomatic Two Steps in Latin America Trip," *The New York Times*, March 25.

Cooksey, Brian. 2004. "Tanzania: Can PRS Succeed Where SAP has Failed?" *Haki Elimu Working Paper Series* 2004, 04.3.

Cox, Robert W. 1997. "A Perspective on Globalization," in Mittelman, J. H., ed., *Globalization: Critical Reflections*. Boulder, CO: Lynne Rienner.

De la Barra, Ximena and Richard A. Dello Buono. 2008. *Latin America After the Neoliberal Debacle*. Lanham, MD: Rowman and Littlefield.

Dollar, David. 2001. "Globalization, Inequality, and Poverty Since 1980." Development Research Group. Washington, DC: World Bank.

Eckstein, Susan (ed.). 2001. *Power and Popular Protest: Latin American Social Movements*. Berkeley: University of California Press.

Economic Commission for Africa. 2007. *Economic Report on Africa, 2007: Accelerating Africa's Development Through Diversification*. Addis Ababa: ECA.

Ellis, Frank and Ntengua Mdoe. 2003. "Livelihoods and Rural Poverty Reduction in Tanzania." *World Development* 31, 8.

Erdal, Semra. 2005. "Perspectives on Poverty and Agriculture in Rural Tanzania," A Working Paper.

Eriksson, Jenny. 2004. "Restricted Possibilities of Unionization Within the Maquila Industry of El Salvador," Masters Thesis, University of Lund, School of Social Work SOL 061, Sweden, Spring Semester 2004. www.forumsyd.org/templates/FS_Article Type A. aspx?id=5378–35k

Friedman, Thomas L. 2007. "It is a Flat World After All," in Roberts, J. T. and A. B. Hite, eds., *The Globalization and Development Reader: Perspectives on Development and Global Change*. Malden, MA: Blackwell.

Ghana Country Economic Memorandum (CEM). 2007. A Summary of the Key Findings presented at the high-level workshop on growth and poverty reduction in Accra, Ghana. Washington, DC: World Bank.

Goulet, Dennis. 2002. "Inequalities in the Light of Globalization," *Kruoc Institute Occasional Paper* #22:OP:2.

Hytrek, Gary and Kristine M. Zentgraf. 2008. *America Transformed: Globalization, Inequality and Power*. New York: Oxford University Press.

International Monetary Fund. 2008a. *United Republic of Tanzania: Poverty Reduction Strategy Paper. Annual Implementation Report 2006/7*. Washington, DC: IMF.

———. 2008b. *Tanzania: Country Report*. Washington, DC: IMF.

Japan Bank for International Cooperation. 2006. *2006 Report*. Tokyo: JBIC.

Kaufman, Leslie and David Gonzalez. 2001. "Labor Standards Clash With Global Reality," *The New York Times*, April 24.

Kayizzi-Mugerwa, Steve. 2001. "Globalization, Growth and Income Inequality: The Africa Experience." OECD Development Center, Working Paper No 186.

Khor, Martin. 2006. "The Impact of Globalization and Liberalization on Agriculture and Small Farmers in Developing Countries: The Experience of Ghana," *Developing Network* (TWN).

Kirkbride, Paul (ed.). 2001. *Globalization: The External Pressures*. Chichester: John Wiley and Sons, Ltd.

Kuteesa, Florence N. and Rosetti Nabbumba. 2004. "HIPC Debt Relief and Poverty Reduction Strategies: Uganda's Experience." www.fondad.org

Lancaster, Carol. 1991. "Economic Reform in Africa: Is it Working?" in Olusegun Obasanjo and Haus d'Orville, eds., *The Leadership Challenge of Economic Reforms in Africa*. New York: Crane Rusk.

Lerman, Robert I. 2002. "Globalization and the Fight Against Poverty," http://www.urban.org/url.cfm.

Magbadelo, John Olushola. 2005. "Westernism, Americanism, Globalism and Africa's Marginality." *Journal of Third World Studies* XXII, 2.

Mainwaring, Scott and Terrence R. Scully (eds.). 2003. *Christian Democracy in Latin America: Electoral Competition and Regime Conflicts*. Stanford: Stanford University Press.

Makoba, Johnson W. 1998. *Government Policy and Public Enterprise Performance in Sub-Saharan Africa: The Case Studies of Tanzania and Zambia, 1964–1984*. Lewiston, NY: Edwin Mellen.

Martin, Xavier S. 2002. "The Disturbing 'Rise' of Global Income Inequality." *National Bureau of Economic Research*, Cambridge, Massachusetts, Working Paper No. 8904.

Mbelle, A. V. Y. 2007. "Financing Development and Poverty Reduction in East Africa," paper presented at the 11th Intergovernmental Committee of Experts (ICE), Bujumbura, Burundi, April 16–20, 2007.

McMichael, Philip. 2008. *Development and Social Change: A Global Perspective*. Fourth Edition. Thousand Oaks, CA: Pine Forge.

———. 2000. *Development and Social Change: A Global Perspective*. Thousand Oaks, CA: Pine Forge.

Mittelman, James H. (ed.). 1997. *Globalization: Critical Reflections*. Boulder, CO: Lynne Rienner.

Nissanke, Machiko and Erik Thorbecke. 2006. "Channels and Policy Debate in the Globalization-Inequality-Poverty Nexus," *World Development*, 34: 8.

Pakulski, Jan. 2004. *Globalizing Inequalities: New Patterns of Social Privelege and Disadvantage*. Crows Nest, Australia: Allen and Unwin.

Philip, Gulley. 2003. *Democracy in Latin America: Surviving Conflict and Crisis?* Cambridge, UK: Polity.

Robbins, Richard H. 2002. *Global Problems and the Culture of Capitalism*. Boston: Allyn and Bacon.

Roberts, J. Timmons and Amy B. Hite (eds.). 2007. *The Globalization and Development Reader: Perspectives on Development and Global Change*. Malden, MA: Blackwell.

Rugumamu, Severine. M. 2001. "Globalization and Africa's Future: Towards Structural Stability, Integration and Sustainable Development." African Association of Political Science, *Occasional Paper Series* 5, 2: 1–92.

Sacks, Jeffrey D. 2007. "The Antiglobalization Movement," in Roberts, J. T. and A. B. Hite (eds.), *The Globalization and Development Reader: Perspectives on Development and Global Change*. Malden, MA: Blackwell.

Sassen, Saskia. 1998. *Globalization and Its Discontents*. New York: The New Press.

Sharer, Robert L., Hema R. DeZoysa, and Calvin A. McDonalds (eds.) 1995. "Uganda: Adjustment with Growth, 1987–1994." Washington, DC: International Monetary Fund.

Sklair, Leslie. 1999. *Sociology of the Global System*. Baltimore: The Johns Hopkins University Press.

Ssewanyana, N. S., A. J. Okidi, D. Angemi, and V. Barungi. 2004. "Understanding Determinants of Income Inequality in Uganda." http://www.bepress.com/case/Paper229.

Stein, Howard. 1992. "Deindustrialization, Adjustment, the World Bank and the IMF in Africa," *World Development* 20, 7.

Stiglitz, Joseph E. 2007. "Globalism's Discontents," in Roberts, J. T. and A. B. Hite (eds.), *The Globalization and Development Reader: Perspectives on Development and Global Change*. Malden, MA: Blackwell.

Tanzania Social and Economic Trust. 2006. *Tanzania after Poverty Reduction and Growth Facility (PRGF): A New Role for the IMF*, Briefing Paper, April 14, 2006.

Thomas, Caroline and Peter Wilkin (eds.). 1997. *Globalization and the South*. New York: St. Martin's.

Vanden, Harry E. and Gary Prevost. 2006. *Politics of Latin America: The Power Game*. New York: Oxford University Press.

Weiner, Tim. 2002. "Monterey's Poor Sinking in Rising Economic Tide," *The New York Times*, March 21.

Weisman, Robert. 2003. "Grotesque Inequality: Corporate Globalization and the Global Gap Between Rich and Poor," http://www.thirdworldtraveler.com.

World Bank. 2007. *The Global Citizen's Handbook: Facing Our World's Crises and Challenges*. New York: Harper Collins.

Worrell, Rodney. 2001. "Whither Global Africa? A Case of Pan-Africanism," *Africa Quarterly* 41, 1–2.

6

Global Capitalism in Crisis: Globalization, Imperialism, and Class Struggle

Berch Berberoglu

Global capitalism is in serious crisis, and the current global economic recession is the worst economic downturn since the Great Depression of the early twentieth century.[1] As neoliberal capitalist globalization comes under mounting criticism and attack across the world, and as the current deepening global economic crisis takes on depression-era characteristics, neoliberalism and neoliberal economic policies have now become thoroughly discredited in many countries around the globe. As millions of unemployed working people look for a job to pay for their basic necessities, capitalist states throughout the world have been spending hundreds of billions of tax dollars to bail out failed commercial and financial institutions, with more than a trillion dollars of economic stimulus program by the United States government alone and several hundreds of billions of dollars by other governments in Europe, China, and elsewhere, to save the global capitalist system from total collapse.

Despite the active role of the imperial state in intervening in the global capitalist economy to reverse its decline and fall, corporations and banks, ranging from mainstays of capitalist economies, such as General Motors and Chrysler, to some of the biggest commercial banks, such as Citi Group and Bank of America, to financial and brokerage firms, insurance companies, and real estate underwriters, such as Lehman Brothers, American Insurance Group (AIG), Fannie Mae and Freddie Mack, have come to a halt and are on the verge of bankruptcy, threatening to take down with them the entire global capitalist system. As a result, and with a ripple effect across the U.S. economy over a period of less than a year, the DOW has plunged more than 50 percent from its highs of 14,000 in late 2007 to below 7,000 in early 2009, with more than a trillion dollars of value lost in the stock market—a development that has shaken markets across the globe and resulted in similar losses in stock markets throughout the capitalist world. Clearly, capitalism is going through its biggest worldwide economic decline since the Great Depression of 1929, and this signals the end of global capitalism as we have come to know it.

Given the failure of neoliberalism and neoliberal capitalist globalization, and as an extension of the capitalist system in general in the early twenty-first century, many are now asking what is in store for the future of the global economy and which direction it will take in the period ahead. This was the main topic of discussion among the leaders of the world's leading economies at the G-20 meetings in London in April 2009, which resulted in guarded optimism that through substantial reforms in global financial institutions and an active interventionist state that monitors the situation with greater regulation of the economy, the evolving economic situation might provide the basis of a new global economic order. What that order will look like and what role the United States will play in it are questions that remain open and contingent on the solutions adopted at the national and global levels, especially in Europe, East Asia, and other emergent centers of global economic power in the aftermath of the current global economic crisis.[2]

In this chapter, I examine the relationship between globalization and imperialism, the dynamics, contradictions, and crisis of global capitalism, and its political-military arm the imperial state, the developing and maturing class struggle, and the prospects for social change and transformation of global capitalism. I examine these within the context of the globalization of capital in the late twentieth and early twenty-first centuries and map out the political implications of this process for the future course of capitalist development on a world scale.

To understand fully the nature, dynamics, and contradictions of the current global economic crisis, we must first examine the theoretical and historical underpinnings of neoliberal globalization and its relationship to capitalist imperialism.

Globalization: Theoretical and Historical Context

The globalization of capital—the accumulation of capital from the national to the international level, where the worldwide operations of the transnational corporations have led to the rise of vast capitalist empires across the world—had developed long before the concept of "globalization" became fashionable among Western intellectuals during the closing decades of the twentieth and the turn of the twenty-first centuries.

Writing in the early twentieth century, John A. Hobson, a liberal British political economist and member of Parliament, was among the very first critics of British imperialism who in his book *Imperialism: A Study*, published in 1905, pointed out in no uncertain terms the very essence of the global expansion of capital and the domination of the global economy by capitalist interests that defined the nature and dynamics of international economic relations and, by extension, political relations of control and domination of the world by powerful financial interests (Hobson [1905] 1972). "Imperialism," wrote Hobson, "implies the use of the machinery of government by private interests, mainly capitalist, to secure for them economic gains outside their country" (1972, 94). He went on to state, "The economic root of Imperialism is the desire of strong organized industrial and financial interests to secure and develop at the public expense and by the public force private markets for their surplus goods and their surplus capital" (Hobson 1972, 106). "The growing cosmopolitanism of capital," he added, "has been the greatest economic change of recent generations. Every advanced industrial nation has been tending to place a larger share of its capital outside the limits of its own political area, in foreign countries, in colonies, and to draw a growing income from this source" (Hobson 1972, 51). Thus, "aggressive Imperialism...is a source of great gain to the investor who cannot find at home the profitable use he seeks for his capital and insists that his government should help him to profitable and secure investments abroad" (Hobson 1972, 55).

Extending Hobson's analysis of British imperialism to the rest of the capitalist world and placing it in historical context and in class terms, V. I. Lenin in his book *Imperialism: The Highest Stage of Capitalism*, published in 1917, developed a general Marxist theory of capitalist imperialism, viewing it as an extension of the logic of capital accumulation and capitalist development on a world scale (Lenin [1917] 1975). He pointed out that capitalism in its highest and most mature monopoly stage has spread to every corner of the world and thus has planted the seeds of its own contradictions everywhere (Lenin 1975, 699–700).

The beginning point of Lenin's analysis of imperialism is his conception of the dynamics of modern capitalism: the concentration and centralization of production. "The enormous growth of industry and the remarkably rapid concentration of production in ever-larger enterprises," he wrote, "are the most characteristic features of capitalism" (Lenin 1975, 642). Moreover, "at a certain stage of its development," he added, "concentration itself, as it were, leads straight to monopoly" (Lenin 1975, 643). "Today [in 1916]," he concluded, "monopoly has become a fact...and that the rise of monopolies, as a result of the concentration of production, is a general and fundamental law of the present stage of development of capitalism" (Lenin 1975, 645).

The underlying argument in Lenin's analysis of imperialism as the highest stage of capitalism is that imperialism is the necessary outcome of the development of capitalism:

> Imperialism emerged as the development and direct continuation of the fundamental characteristics of capitalism in general. But capitalism only became capitalist imperialism at a definite and very high stage of its development....Economically, the main thing in this process is the displacement of capitalist free competition by capitalist monopoly....Monopoly is the transition from capitalism to a higher system. If it were necessary to give the briefest possible definition of imperialism we should have to say that imperialism is the monopoly stage of capitalism. (Lenin 1975, 700)

Thus, in summarizing the fundamental features of imperialism (or monopoly capitalism operating on a world scale), Lenin concluded, "Imperialism is capitalism in that stage of development in which the dominance of monopolies and finance capital is established; in which the export of capital has acquired pronounced importance; in which the division of the world among the international trusts has begun; in which the division of all territories of the globe among the biggest capitalist powers has been completed" (1975, 700).

Lenin's emphasis on the importance of the export of capital is crucial from the angle of its implications concerning the transformation of relations of production abroad. With the export of capital as the primary source of the globalization of capital and capitalist class relations on a world scale, capitalism effected transformations in the class structure of societies with which it came into contact. As a result, the class contradictions of the capitalist mode of production became the outcome of the dominant form of exploitation of labor through the instrumentality of imperialist expansion throughout the world. It is in this context of the developing worldwide contradictions of advanced, monopoly capitalism that Lenin pointed out, "[I]mperialism is the eve of the social revolution of the proletariat...on a worldwide scale" (1975, 640).

Imperialism and Globalization

Imperialism is the highest stage of capitalism operating on a world scale, and globalization is the highest stage of imperialism that has penetrated every corner of the world. Both are an outgrowth of twentieth-century monopoly capitalism—an inevitable consequence, or manifestation, of monopoly capital that now dominates the world capitalist political economy. Thus, the current wave of neoliberal globalization is an extension of this process that operates at a more advanced and accelerated level.

A central feature of this current phase of transnational capitalism, besides its speed and intensity, is the increased privatization of various spheres of the economy and society. This has especially been the case in areas such as communications, information technology, education, and the cultural sphere, where privatization is becoming increasingly prevalent.

The rate at which these changes have been taking place, and the vigor with which transnational capital has been exercising more power vis-à-vis the state, has led some to declare globalization a qualitatively new stage in the development of world capitalism (Ross and Trachte 1990). However, I argue that these quantitative, surface manifestations of contemporary capitalism, no matter how pervasive they are, do *not* change the fundamental nature of capitalism and capitalist relations, or the nature of the capitalist/imperialist state and the class contradictions generated by these relations, which are inherent characteristics of the system itself. They cannot change the nature of capitalism in any qualitative sense to warrant globalization a distinct status that these critics have come to assign as something fundamentally different than what Marxist political economists have always argued to be the "normal" operation and evolution of global capitalism in the age of imperialism (Szymanski 1981; Warren 1980; Beams 1998; Foster 2002; Harvey 2003).

Today, in the early twenty-first century, the dominant institution that has facilitated global capitalist expansion on behalf of the current center of world imperialism since the post–World War II period—the United States—is the *transnational corporation*. As other capitalist rivals from Europe and the Pacific Basin have recently begun to emerge on the world scene as serious contenders for global economic power, they too have developed and unleashed their own transnational corporate and financial institutions to carve out greater profits, accumulate greater wealth, and thereby dominate the global economy. The transnational corporations and banks, based in the leading centers of world capitalism, have thus become the chief instruments of global capitalist expansion and capital accumulation (Waters 1995; Mittelman and Othman 2002; see also Barnet and Cavenagh 1994; Petras and Veltmeyer 2007). It is, therefore, in the export of capital and its expanded reproduction abroad to accumulate greater wealth for the capitalist classes of the advanced capitalist countries that one can find the motive force of imperialism and capitalist globalization in the late twentieth and early twenty-first centuries.

Globalization today, much as during the earlier stages of capitalism, is driven by the logic of *profit* for the private accumulation of capital based on the exploitation of labor throughout the world. It is, in essence, the highest and most pervasive phase of transnational capitalism operating on a world scale. It is the most widespread and penetrating manifestation of modern capitalist imperialism in the age of the Internet—a development that signifies not only the most thorough economic domination of the world by the biggest capitalist monopolies, but also increasingly direct military intervention by the chief imperialist state to secure the global economic position of its own corporations.[3]

The relationship between the owners of the transnational corporations—the monopoly capitalist class—and the imperialist state and the role and functions of this state, including the use of military force to advance the interests of the monopoly capitalist class, thus reveals the class nature of the imperialist state and the class logic of imperialism and globalization (Warren 1980; Szymanski 1981; Berberoglu 1987, 1992b, 2003). But this logic is more pervasive and is based on a more fundamental class relation between labor and capital that now operates on a global level, that is, a relation based on exploitation. Thus, in the age of globalization, that is, in the epoch of capitalist imperialism, social classes and class struggles are a product of the logic of the global capitalist system based on the exploitation of labor worldwide (Gerstein 1977; Petras 1978; Berberoglu 2009).

Capitalist expansion on a world scale at this stage of the globalization of capital and capitalist production has brought with it the globalization of the production process and the exploitation of wage-labor on a world scale. With the intensified exploitation of the working class at super-low wages in repressive neocolonial states throughout the Third World, the transnational corporations of the leading capitalist states have come to amass great fortunes that they have used to build up a global empire through the powers of the imperial state, which has not hesitated to use its military power to protect and advance the interests of capital in every corner of the globe. It is in this context that we see the coalescence of the interests of the global economy and empire as manifested in control of cheap labor, new markets, and vital sources of raw materials, such as oil, and the intervention of the state to protect these when their continued supply to the imperial center are threatened (Petras and Veltmeyer 2001, 2007).

Imperialism has been an enormous source of profit and wealth for the capitalist class of the advanced capitalist countries, who, through the mechanisms of the transnational monopolies and the imperial state, have accumulated great fortunes from the exploitation of labor on a world scale. Given the uneven development of capitalism, however, some countries have grown more rapidly than others, while previously less developed countries have emerged as new centers of world capitalism. The rivalry between the capitalist classes of the old and newly emergent capitalist states has turned into rivalry among the leading countries within the world capitalist system. This has led to intense competition and conflict between the rising capitalist powers and the declining imperial centers on a world scale, hence leading to shifts in centers of global economic and political power (Pieterse 2004).

The process of global capitalist expansion discussed earlier has produced a number of major consequences, which are examined at length in this chapter. These can be listed briefly as follows:

1. The globalization of capital and the development of capitalism and capitalist relations of production in the less developed capitalist countries resulting in the super-exploitation of a growing working class;
2. The rise of new capitalist centers on the world scene (e.g., Japan, Germany, and the European Union), and other emerging economies (e.g., China and India), thus leading to global rivalry;
3. The necessity to protect and police the empire, hence the procurement and maintenance of a large number of military bases around the world, frequent military intervention in the Third World, and, as a result, an enormous increase in military spending;
4. Economic crisis, decline of the domestic economy, and a reduction in the living standard of U.S. workers, leading to increased class polarization in the advanced capitalist centers;
5. The class contradictions of imperialism and capitalist development on a world scale, preparing the material conditions for intensified class struggles that lead to revolutionary social transformations throughout the world, including the empire's home base.

Focusing on the U.S. experience, it is clear that in the post–World War II period the United States emerged as the dominant power in the capitalist world. In subsequent decades, U.S.-controlled transnational production reached a decisive stage, necessitating the restructuring of the international division of labor, as the export of productive capital brought about a shift in the nature and location of production: the expansion of manufacturing industry on an unprecedented scale into previously pre-capitalist, peripheral areas of the global capitalist economy. This marked a turning point in the rise to world prominence of the U.S. economy and the emergence of the United States as the leading capitalist/imperialist power in the world (Petras and Veltmeyer 2001, 2007; Pieterse 2004).

Although the large-scale U.S. postwar global expansion ushered in a period of unquestioned U.S. supremacy over the world economy and polity during the 1950s and 1960s, the economic strength of U.S. capital over foreign markets through investment, production, and trade during the 1970s took on a new significance—one resulting from the restructuring of the international division of labor. U.S. transnational capital, in line with its transfer of large segments of the production process to the periphery, poured massive amounts of capital into select areas of the Third World, as well as into its traditional bases of foreign investment—Canada and Western Europe—and became the leading center of world capitalism in a new way, that is, by becoming the dominant force in the worldwide production process. Thus, not only did overall U.S. direct investment expand immensely during this period, but also a shift in the form of investment in favor of manufacturing came to constitute the new basis of changes in the international division of labor with great impact on the national economies of both the periphery and the center states, including the United States. This process further fueled the contradictions and conflicts inherent in capitalist production and class relations on a global scale, including inter-imperialist rivalry between the chief capitalist states, on the one hand, and the exploitation of labor on a global scale, on the other, with all the consequences associated with this process—a process that has led to the crisis of global capitalism.

Contradictions and Crisis of Global Capitalism and the Imperial State

The development of capitalism over the past hundred years has formed and transformed capitalism in a crucial way, one that is characterized by periodic crises resulting from the capitalist business cycle that now unfolds at the global level. The current crisis of global capitalism is an outcome of the consolidation of monopoly power that the globalization of capital has secured for the transnational monopolies (Sassen 2009). This has led to a string of problems associated with the contradiction between the expanded forces of production and existing exploitative social relations of production (i.e., class relations), which manifests itself in a number of ways, including:

1. The problem of overproduction, resulting from the imbalance created between wages and prices of commodities fueled by low purchasing power;
2. Sub-prime mortgage and credit card debt and rising foreclosures and bankruptcies as the unemployed become unable to pay off their debts;
3. Increasing unemployment and underemployment resulting from outsourcing of jobs to low-wage sweatshops in export processing zones abroad, compounded by the continued application of technology in production (i.e., automation);
4. Intensification of the exploitation of labor through expanded production and reproduction of surplus value and profits by further accumulation of capital and the reproduction of capitalist relations of production on a world scale;
5. Increased polarization of wealth and income at the national and global levels between the capitalist and working classes and growth in numbers of the poor and marginalized segments of the population throughout the world.

These and other related contradictions and crises of global capitalism define the parameters of modern capitalist imperialism and provide us the framework of discussion on the nature and dynamics of imperialism and globalization in the world today.

Given the logic of global capital accumulation in late capitalist society, it is no accident that the decline of the domestic economy of advanced capitalist countries over the past three decades corresponds to the accelerated export of capital abroad in search of cheap labor, access to raw materials, new markets, and higher rates of profit. The resulting deindustrialization of the domestic economy has had a serious impact on workers and other affected segments of the laboring population and has brought about a major dislocation of the national economy (Phillips 1998; Berberoglu 2003).[4] This has necessitated increased state intervention on behalf of the monopolies and has heightened the contradictions that led to the crisis of advanced capitalist society in the early twenty-first century.

The widening gap between the accumulated wealth of the capitalist class and the declining incomes of workers (within a deteriorating

national economy and the state's budgetary crisis) has led to the ensuing political crisis within the state apparatus and has sharpened the
class struggle in a new political direction. As the crisis of the capitalist economy has brought the advanced capitalist/imperial state to the
center stage of economic life and revealed its direct ties to the monopolies, thus exacerbating the state's legitimization crisis, the struggles
of the working class and the masses in general are becoming directed
not merely against capital, but against the state itself (Beams 1998).

The crisis of the capitalist state on the global scene is a manifestation of the contradictions of the world economy, which in the early
twenty-first century has reached a critical stage in its development.
The massive flow of U.S. transnational investment throughout the
world, especially in Western Europe, Japan, and other advanced
capitalist regions, has led to the post–World War II reemergence of
inter-imperialist rivalry between the major capitalist powers, while
fostering antagonisms between them in the scramble for the peripheral regions of the global capitalist economy—Latin America, Asia,
Africa, and the Middle East (Hart 1992; Falk 1999; Halliday 2001).

With the integration of the economies of Western Europe into the
European Union (EU) and the emergence of Japan as a powerful economic force in the late twentieth century, the position of the United
States in the global economy has declined relative to both its own
postwar supremacy in the 1940s and 1950s and to other advanced
capitalist economies since that time. Despite the fact that U.S. capital continues to control the biggest share of overseas markets and
accounts for the largest volume of international investments, its hold
on the global economy has recently begun slipping in a manner similar to Britain's in the early twentieth century. This has, in turn, led
the U.S. state to play a more aggressive role in foreign policy to protect U.S. transnational interests abroad. Its massive deployment in the
Middle East in the early 1990s, which led to the Persian Gulf War of
1991, and subsequently its intervention in Afghanistan in 2001 and
war against Iraq in 2003, has resulted in great military expenditures
that translated into an enormous burden on working people of the
United States, who have come to shoulder the colossal cost of maintaining a global empire whose vast military machine encompasses the
world (Berberoglu 2003, 2005).

In the current phase of the crisis of the U.S. economy and the imperial state, the problems the state faces are of such magnitude that they
threaten the supremacy of the United States in the global political
economy and by extension the global capitalist system itself. Internal

economic and budgetary problems have been compounded by ever-growing military spending propped up by armed intervention in the Third World (Iraq, Afghanistan, etc.), while a declining economic base at home manifested in the housing and banking crisis, deindustrialization, and a recessionary economy further complicated by the global rivalry between the major capitalist powers that is not always restricted to the economic field, but has political (and even military) implications that are global in magnitude (Beams 1998; Harvey 2003; see also Panitch and Leys 2003).

The growing prospects of inter-imperialist rivalry between the major capitalist powers, backed up by their states, are effecting changes in their relations that render the global political economy an increasingly unstable character. Competition between the United States, Japan, and European imperial states, and the emergence of China, Russia, and other rival states, are leading them on a collision course for world supremacy, manifested in struggles for markets, raw materials, and spheres of influence in geopolitical—as well as economic—terms, which may in fact lead to a new balance of forces, and, consequently, alliances that will have serious political implications in global power politics. As the continuing economic ascendance of the major capitalist rivals of the United States take their prominent position in the global economy, pressures will build toward the politicization and militarization of these states from within, where the forces of the leading class bent on dominating the world economy will press forward with the necessary political and military corollary of their growing economic power in the global capitalist system (Hart 1992; Falk 1999), as has been the case with the German, French, Russian, and Chinese opposition to war against Iraq in the U.N. Security Council in 2003.

These developments in global economic and geopolitical shifts in the balance of forces among the major powers will bring to the fore new and yet untested international alliances for world supremacy and domination in the post–cold war era. Such alliances will bring key powers such as Russia and China into play in a new and complicated relationship that holds the key for the success or failure of the new rising imperial centers that will emerge as the decisive forces in the global economic, political, and military equation in the early decades of the twenty-first century (Halliday 2001; Guthrie 2006; Stephens 2009).

The contradictions and conflicts imbedded in relations between the rival states of the globe will again surface as an important component of international relations in the years ahead. And these are part and parcel of the restructuring of the international division of labor and the transfer of production to overseas territories in line with the globalization of capital on a worldwide basis—a process that has serious consequences for the economies of both the advanced capitalist and less developed capitalist countries. Economic decline in the imperial centers (manifested in plant closings, unemployment, and recession) and super-exploitation of workers in the Third World (maintained by repressive military regimes) yield the same combined result that has a singular global logic: the accumulation of transnational profits for the capitalist class of the advanced capitalist countries—above all, that of the United States, the current center of global capitalism. It is in this context of the changes that are taking place on a world scale that the imperial state is beginning to confront the current crisis of global capitalism.

The contradictions of the unfolding process of global expansion and accumulation have brought to the fore new political realities: renewed repression at home and abroad to control an increasingly frustrated working class in the imperial heartland, and a militant and revolutionary mass of workers and peasants in the neocolonial states of the Third World poised to resist capitalist globalization (Houtart and Polet 2001). It is these inherent contradictions of modern monopoly capital that are making it increasingly difficult for the imperial state to control and manage the global political economy, while at the same time preparing the conditions for international solidarity of workers in confronting global capital on a world scale.

Imperialism, Globalization, and Class Struggle

The global expansion of capital has had varied effects in the international and domestic economic spheres. At the global level, it has meant first and foremost the ever-growing exploitation of workers through the use of cheap labor. In addition, it has caused a depletion of resources that could be used for national development, environmental pollution, and other health hazards; a growing national debt tying many countries to the World Bank, the International Monetary Fund, and other imperialist financial institutions; and a growing militarization of society through the institution of brutal military and civilian dictatorships that violate basic human rights. The domination and control of Third World countries for transnational profits through the instrumentality of the imperial state has at the same time created various forms of dependence on the center that has become a defining characteristic of globalization and imperialism (Amaladoss 1999; Sklair 2002).

Domestically, the globalization of capital and imperialist expansion has had immense dislocations in the national economies of imperialist states. Expansion of manufacturing industry abroad has meant a decline in local industry, as plant closings in the United States and other advanced capitalist countries have worsened the unemployment situation. The massive expansion of capital abroad has resulted in hundreds of factory shutdowns with millions of workers losing their jobs, hence the surge in unemployment in the United States and other imperialist states (Wagner 2000). This has led to a decline in wages of workers in the advanced capitalist centers, as low wages abroad have played a competitive role in keeping wages down in the imperialist heartlands. The drop in incomes among a growing section of the working class has thus lowered the standard of living in general and led to a further polarization between labor and capital (Berberoglu 1992a; 2002).

The dialectics of global capitalist expansion, which has caused so much exploitation, oppression, and misery for the peoples of the world, both in the Third World and in the imperialist countries themselves, has in turn created the conditions for its own destruction. Economically, it has afflicted the system with recessions, depressions, and an associated realization crisis; politically, it has set into motion an imperial interventionist state that through its presence in every corner of the world has incurred an enormous military expenditure to maintain an empire, while gaining the resentment of millions of people across the globe who are engaged in active struggle against it.[5]

The imperial capitalist state, acting as the repressive arm of global capital and extending its rule across vast territories, has dwarfed the militaristic adventures of past empires many times over. The global capitalist state, through its political and military supremacy, has come to exert its control over many countries and facilitate the exploitation of labor on a world scale. As a result, it has reinforced the domination of capital over labor and its rule on behalf of capital. This, in turn, has greatly politicized the struggle between labor and capital and called for the recognition of the importance of political organization that many find necessary to effect change in order to transform the capitalist-imperialist system.

Understanding the necessity of organizing labor and the importance of political leadership in this struggle, radical labor organizations have in fact taken steps emphasizing the necessity for the working class to mobilize its ranks and take united action to wage battle against capitalist imperialism globally. In this sense, labor internationalism (or the political alliance of workers across national boundaries in their struggle against global capitalism) is increasingly being seen as a political weapon that would serve as a unifying force in labor's frontal attack on capital in the ensuing class struggle (Beams 1998).[6]

Imperialism today represents a dual, contradictory development whose dialectical resolution is an outcome of its very nature—a product of its growth and expansion across time and space within the confines of a structure that promotes its own destruction and demise. However, while the process itself is a self-destructing one, it is important to understand that the nature of the class struggle that these contradictions generate is such that the critical factor that tips the balance of class forces in favor of the proletariat to win state power is political organization, the building of class alliances among the oppressed and exploited classes, the development of strong and theoretically well-informed revolutionary leadership that is organically linked to the working class, and a clear understanding of the forces at work in the class struggle, including, especially, the role of the state and its military and police apparatus—the focal point of the struggle for state power (Szymanski 1978; Berberoglu 2001; Knapp and Spector 1991). The success of the working class and its revolutionary leadership in confronting the power of the state thus becomes the critical element ensuring that, once captured, the state can become an instrument that the workers can use to establish their rule and in the process transform society and the state itself to promote proletarian interests in line with its vision for a new society free of exploitation and oppression, one based on the rule of the working class and the laboring masses in general.

Our understanding of the necessity for change and social transformation, which is political in nature, necessitates a clear, scientific understanding of modern imperialism in its late twentieth- and early twenty-first-century form so that this knowledge can be put to use to facilitate the class struggle in a revolutionary direction. In this context, one will want to know not only the extent and depth of global capitalist expansion, but also its base of support, its linkage to the major institutions of capitalist society (above all the state, but also other religious, cultural, and social institutions), the extent of its ideological hegemony and control over mass consciousness, and other aspects of social, economic, political, and ideological domination. Moreover—and this is the most important point—one must study its weaknesses, its problem areas, its vulnerabilities, its weak links, and the various dimensions of its crisis—especially those that affect its continued reproduction and survival. Armed with this knowledge, one would be better equipped to confront capital and the capitalist states in the struggle to transform imperialism and the globalization process that today, in the early twenty-first century, has come to dominate the world.

Class Struggle and Transformation of Global Capitalism

Thanks to the growing literature on globalization and global capitalism, we now have a greater understanding of the structure of capitalist imperialism and its contradictions. We know, for example, the extent of global capitalist expansion, the nature of imperialist intervention around the world, and the various social, economic, and political contradictions of imperialism today (Szyamnski 1981; Berberoglu 1987, 2003; Petras and Veltmeyer 2001).

The question that one now confronts is a *political* one. Given what we know of imperialism and its class contradictions on a world scale, how will the peoples' movements respond to imperialism *politically* worldwide? What strategy and tactics will be adopted to confront this colossal force? It is important to think about these questions concretely, in a practical way—one that involves a concrete scientific analysis and organized political action.

One central location of this battleground has been the Third World, where efforts toward the development of solidarity among workers to build the basis of a true labor internationalism have been quite successful. Armed with proletarian solidarity, a rank-and-file international workers' movement mobilized across national boundaries has the potential to play a strong role in bringing together workers from various countries in their struggle against transnational capital and the global capitalist system. Such international solidarity among Third World workers could represent a mighty force in the struggle against imperialism and capitalist exploitation throughout the world (Bina and Davis 2002).

Strikes, demonstrations, and mass protests initiated by workers and other popular forces have become frequent in a growing number of countries controlled by imperialism in recent years. Working people are rising up against the local ruling classes, the state, and the transnational monopolies that have together effected the superexploitation of labor for decades. Varied forms of class struggle, on the one hand, and the struggle for national liberation led by labor, on the other, are two sides of the same process of struggle for the transformation of society now underway in many countries under the grip of foreign capital.

The logic of transnational capitalist expansion on a global scale is such that it leads to the emergence and development of forces in conflict with this expansion. The working class has been in the forefront of these forces. And strikes, mass demonstrations, political protest, confrontation with the local client state machine, armed insurrection, civil war, and revolutionary upheavals are all part and parcel of the contradictory nature of relations imposed on the laboring people by imperialism and its client states throughout the Third World.

Another important location of this battleground is Europe. The influence of the European Union, led by German imperialism, is growing and expanding throughout the world. The danger stemming from this economic expansion is real, and such expansion will increasingly take a political and military form to protect this expanding economic interest worldwide. However, the growing German influence in Europe has become a focal point of resistance against German and European imperialism throughout the continent as part of the effort directed at confronting the forces of globalization on the European mainland.

Yet another rallying point of struggle has been around the North American Free Trade Agreement (NAFTA) and other neoliberal policies of U.S. imperialism in Mexico and within the United States itself. Here, it is important to note the protracted battle that has been waged by Zapatista National Liberation Army in Chiapas against U.S. imperialism for its economic intervention in Mexico and elsewhere to dominate both the North American and Latin American economies through this agreement. This has been an important effort on the part of labor to build solidarity between U.S. and Mexican (and other Latin American) labor, progressive trade unions, and leftist political organizations in building links and alliances that can translate into concrete political action, including general strikes, demonstrations, and protests along the U.S.-Mexican border—actions that represents the unified efforts of both U.S. and Mexican workers in confronting Mexican capital and the Mexican state, as well as U.S. transnationals and the U.S. imperial state (Bacon 2004).

All of these efforts have become important components of a much broader international solidarity of working people that is yet to develop between the workers of the Third World and workers in the advanced capitalist countries in North America, Europe, and elsewhere. Elements of this new emergent solidarity were seen in the recent protests and demonstrations in Seattle (in November 1999), Washington, D.C. (in April 2000), Prague (in September 2000), and several more recent protests in a number of cities in the United States, Canada, and Europe in 2001 and 2002, where labor has played an important role in building the basis of a solidarity across many groups that are allied in this struggle. The most recent demonstrations in France, Italy, Britain, and the United States against the global economic crisis in early 2009 are a continuation of this struggle that is now global. And this alliance and struggle of working people will surely grow and spread further to other parts of the world as well in the coming years (Wallach and Sforza 1999; Starr 2001; Houtart and Polet 2001; Smith and Johnston 2002; Katsiaficas and Yuen 2002; Berberoglu 2005, 2009).

Finally, another important arena of political struggle has been the building of solidarity with the remaining socialist states that have come under imperialist attack. This has included support of movements that are struggling to defeat the reactionary, procapitalist forces in Eastern Europe and the former Soviet Union in order to build a new type of socialist society that is based on the working class and led by the workers themselves.

Together, these struggles have been effective in frustrating the efforts of imperialism to expand and dominate the world, while at the same time building the basis of an international working-class movement that finally overcomes national, ethnic, cultural, and linguistic boundaries that artificially separate the workers in their fight against capitalism and imperialism. The solidarity achieved through this process has helped expand the strength of the international working class and increased its determination to defeat imperialism and all vestiges of global capitalism throughout the world and build a new egalitarian world social order that advances the interests of the working people and ultimately all of humanity.

Notes

1. This chapter is a revised and updated version of chapter 5 of my book *Globalization and Change: The Transformation of Global Capitalism* (Lanham, MD: Lexington Books, 2005). It is published here in its revised form with permission from Rowman and Littlefield.
2. These questions are taken up for closer study by Jan Nederveen Pieterse in chapter 2 of this book.

3. The U.S. invasion and occupation of Iraq in 2003 is a clear example of this, where corporations such as Halliburton, Blackwater, Caci, KBR, Titan, and others have been the direct beneficiaries of this intervention. See the DVD "Iraq for Sale: The War Profiteers" (available on-line through www.amazon.com)

4. This paradox of growth and expansion of capital on a world scale, simultaneously with the decline and contraction of the domestic economy, is a central feature of globalization and imperialism at its highest and most intense stage of worldwide capitalist expansion. See Berberoglu, *Globalization of Capital and the Nation-State.*

5. While one consequence of imperialism and globalization has been economic contraction and an associated class polarization, a more costly and dangerous outcome of this process has been increased militarization and intervention abroad, such that the defense of an expanding capitalist empire worldwide has come to require an increasing military presence and a permanent interventionist foreign policy to keep the world economy clear of obstructions that go against the interests of the transnational monopolies. However, such aggressive military posture has had (and continues to create) major problems for the imperialist state and is increasingly threatening its effectiveness and, in the long run, its very existence.

6. The necessity of the struggle against global capital in an organized political fashion has been emphasized by working-class organizations, and this has led to several successful revolutions during the twentieth century. Throughout this period, working-class organizations have emphasized the centrality of international working-class solidarity (or proletarian internationalism) for any worldwide effort to wage a successful battle against global capitalism.

References

Amaladoss, Michael (ed.). 1999. *Globalization and Its Victims As Seen by Its Victims.* Delhi, India: Vidyajyoti Education and Welfare Society.

Bacon, David. 2004. *The Children of NAFTA: Labor Wars on the U.S./Mexico Border.* Berkeley: University of California Press.

Barnet, Richard and John Cavenagh. 1994. *Global Dreams: Imperial Corporations and the New World Order.* New York: Simon & Schuster.

Beams, Nick. 1998. *The Significance and Implications of Globalization: A Marxist Assessment.* Southfield: Mehring Books.

Berberoglu, Berch. 1987. *The Internationalization of Capital: Imperialism and Capitalist Development on a World Scale.* New York: Praeger.

———. 1992a. *The Legacy of Empire: Economic Decline and Class Polarization in the United States.* New York: Praeger.

———. 1992b. *The Political Economy of Development.* Albany: State University of New York Press.

———. 2001. *Political Sociology: A Comparative/Historical Approach.* 2nd ed. New York: General Hall.

———. (ed.). 2002. *Labor and Capital in the Age of Globalization: The Labor Process and the Changing Nature of Work in the Global Economy.* Lanham, MD: Rowman and Littlefield.

———. 2003. *Globalization of Capital and the Nation-State*. Lanham, MD: Rowman and Littlefield.

———. 2005. *Globalization and Change: The Transformation of Global Capitalism*. Lanham, MD: Lexington Books.

———. 2009. *Class and Class Conflict in the Age of Globalization*. Lanham, MD: Lexington Books.

Bina, Cyrus and Chuck Davis. 2002. "Dynamics of Globalization: Transnational Capital and the International Labor Movement," in Berch Berberoglu, ed., *Labor and Capital in the Age of Globalization*. Lanham, MD: Rowman and Littlefield.

Falk, Richard. 1999. *Predatory Globalization: A Critique*. Malden, MA: Blackwell.

Foster, John B. 2002. AMonopoly Capital and the New Globalization,@ *Monthly Review* 53, 8, January.

Gerstein, Ira. 1977. "Theories of the World Economy and Imperialism," *Insurgent Sociologist* 7, 2, Spring.

Guthrie, Doug. 2006. *China and Globalization: The Social, Economic and Political Transformation of Chinese Society*. New York: Routledge.

Halliday, Fred. 2001. *The World at 2000*. New York: St. Martin=s.

Hart, Jeffrey A. 1992. *Rival Capitalists: International Competitiveness in the United States, Japan, and Western Europe*. Ithaca, NY: Cornell University Press.

Harvey, David. 2003. *The New Imperialism*. New York: Oxford University Press.

Hobson, John A. [1905] 1972. *Imperialism: A Study*. Rev. ed. Ann Arbor: University of Michigan Press.

Houtart, Francois and Francois Polet (eds.). 2001. *The Other Davos Summit: The Globalization of Resistance to the World Economic System*. London: Zed Books.

Katsiaficas, George and Eddie Yuen (eds.). 2002. *The Battle of Seattle: Debating Capitalist Globalization and the WTO*. New York: Soft Skull.

Knapp, Peter and Alan J. Spector. 1991. *Crisis and Change: Basic Questions of Marxist Sociology*. Chicago: Nelson-Hall.

Lenin, V. I. [1917] 1975. Lenin, *Imperialism: The Highest Stage of Capitalism*. In *Selected Works*. Vol. 1. Moscow: Foreign Languages Publishing House.

Mittelman, James H. and Norani Othman (eds.). 2002. *Capturing Globalization*. New York: Routledge.

Panitch, Leo and Colin Leys (eds.). 2003. *The New Imperial Challenge*. New York: Monthly Review Press.

Petras, James. 1978. *Critical Perspectives on Imperialism and Social Class in the Third World*. New York: Monthly Review Press.

Petras, James and Henry Veltmeyer. 2001. *Globalization Unmasked: Imperialism in the 21st Century*. London: Zed Books.

———. 2007. *Multinationals on Trial: Foreign Investment Matters*. Burlington, VT: Ashgate.

Phillips, Brian. 1998. *Global Production and Domestic Decay: Plant Closings in the U.S.* New York: Garland.

Pieterse, J. Nederveen. 2004. *Globalization or Empire?* New York: Routledge.

132 Berch Berberoglu

Ross, Robert J. S. and Kent C. Trachte. 1990. *Global Capitalism: The New Leviathan.* Albany: State University of New York Press.

Sassen, Saskia. 2009. "Too Big To Save: The End of Financial Capitalism." *Open Democracy News Analysis,* April 2.

Sklair, Leslie. 2002. *Globalization: Capitalism and Its Alternatives.* New York: Oxford University Press.

Smith, Jackie G. and Hank Johnston (eds). 2002. *Globalization and Resistance: Transnational Dimensions of Social Movements.* New York: Routledge.

Starr, Amory. 2001. *Naming the Enemy: Anti-corporate Movements Confront Globalization.* London: Zed Books.

Stephens, Philip. 2009. "A Summit Success That Reflects a Different Global Landscape." *Financial Times,* April 3, 9.

Szymanski, Albert J. 1978. *The Capitalist State and the Politics of Class.* Cambridge, MA: Winthrop.

———. 1981. *The Logic of Imperialism.* New York: Praeger.

Wagner, Helmut (ed.). 2000. *Globalization and Unemployment.* New York: Springer.

Wallach, Lori and Michelle Sforza. 1999. *Whose Trade Organization? Corporate Globalization and the Erosion of Democracy.* Washington, DC: Public Citizen.

Warren, Bill. 1980. *Imperialism, Pioneer of Capitalism.* New York: Verso.

Waters, Malcolm. 1995. *Globalization: The Reader.* New York: Routledge.

7

Globalization and China: From Neoliberal Capitalism to State Developmentalism in East Asia

Alvin Y. So

During the cold war era, China was generally seen by the Left in the West as a model of revolutionary socialism. The Left was especially attracted to the Maoist policies of public ownership, egalitarianism, mass mobilization, militant anti-imperialism, and the rejection of a reformist road to socialism (Halliday 1976; Petras 1997). Nevertheless, in the late 1970s, when the advanced capitalist states lowered their hostility toward communist China and welcomed China back to the global economy, China replaced Maoist policies with "market socialism." Since the late 1970s China's economic development has stunned the world. The country has become one of the world's largest exporters of manufactured goods and sites for transnational investments, while purportedly lifting hundreds of millions out of poverty.

In the West, the Left is divided on how to interpret China's recent transformation at the turn of the twenty-first century. Some see China's market socialism offer tremendous opportunities for achieving growth and poverty reduction, and welcome China's regional and global emergence as it could serve as a counterweight to U.S.-driven neoliberal and militarized capitalism (Silver and Arrighi 2000). Others denounce China's recent transformation as moving toward a neoliberal economy that contains the seeds for the reemergence of a foreign capitalist-dominated state (Petras 2006; Burkett and Hart-Landsberg 2005).

In this chapter, I argue that China's recent transformation actually is closer to the East Asian model of the developmental state than to the Western neoliberal model. I first present the distinctive features of Chinese state developmentalism and explain how this model is different from that of neoliberalism. Following this, I trace the transition from neoliberalism to state developmentalism in China over the past two decades. Finally, I discuss the future trajectory of state developmentalism.

Neoliberal Capitalism

In the 1980s, the Chinese state had been faithfully carrying out the policies of neoliberalism in its globalization drive (Harvey 2005). Since the Chinese economy was completely dominated by the state in the Maoist period, the aim of the post-Mao reforms was to reinvent and liberate the market from the state and to reintegrate China into the global capitalist economy in order to speed up capital accumulation. It is with the above neoliberal mind-set that the Chinese state carried out the following policies over the past several decades:

- *Decollectivization.* In the countryside, agricultural communes were dismantled in favor of an individualized "personal responsibility system." Peasant families were given plots of land to cultivate, and they were responsible for their own gains and losses. They were also encouraged to sell their products to rural markets, engage in rural industries, and seek work in nearby township enterprises. Township and village enterprises were created out of the former commune assets, and these became centers of entrepreneurialism, flexible labor practices, and open market competition.
- *Proletarianization of peasants.* At the same time, the loss of collective social rights in the countryside meant the peasants had to face burdensome user charges for schools, medical care, and the like. Forced to seek work elsewhere after the end of collectivism, rural migrants flooded—illegally and without the right of residency—into the cities to form an immense labor reserve (a "floating population" of indeterminate legal status). China is now in the midst of the largest mass migration the world has ever seen (Chan 2003). This rural "floating population" is vulnerable to super-exploitation and puts downward pressure on the wages of urban workers (Pun 1999).
- *Marketization* policy to restore/expand the market. A new labor market was introduced to the Chinese economy in the late 1980s, creating a flexible labor force that is responsive to the ups and downs of the market. After a labor market was set up, the state enterprises were no

longer required to provide lifelong employment and job security to their workers, and were given the autonomy to hire and fire workers in the name of enhancing productivity and efficiency as called upon by neoliberalism.

- *Fiscal Decentralization and the weakening of the central state.* In the mid-1980s, provincial, municipality, county, and township governments were subject to a bottom-up revenue-sharing system that required localities to submit only a portion of the revenues to the upper level, and then they were allowed to retain all, or at least most, of the reminder. This fiscal decentralization policy made local states become independent fiscal entities that had the unprecedented right to use the revenue they retained. As a result, fiscal decentralization had considerably weakened the central state's extractive capacity. The Chinese state was unable to control the extra-budgetary funds of the local governments, and its relative share of tax revenues had decreased to the extent that the Central state has lost effective control over China's economic life (Wang and Hu 2001; Oi 1992).

- *Opening up and spatial differentiation.* There was an open-door policy toward foreign investments. It began with the establishment of four special economic zones (SEZs) in 1979, the opening of fourteen coastal cities and Hainan Island in 1984, and the extension to three delta areas (Pearl River Delta, Yangtze River Delta, and Yellow River Delta) in 1985. The combination of decentralization and opening up to global capital has led to a very uneven pattern of spatial development in China, with rapid economic growth taking place mostly along the eastern coastal subregions. These subregions were characterized by an *"extrovert"* economy, that is, their economies were driven by foreign direct investment and export-led industrialization, and their economic growth relied upon their integration with the global commodity chains. For example, with regard to the commodity chain of athletic shoes, the 1990s observed the trend that transnationals (such as Nike and Reebok) moved their factories from their subcontractors in Taiwan to Guangdong and Fujian. Most of the raw materials were shipped from Taiwan, and the shoe factories in Guangdong were run by Taiwanese resident managers (Chen 2005).

Through the above processes of decollectivization and proletarianization, marketization, fiscal decentralization, opening up, and spatial differentiation, China was moving toward the "neoliberal" capitalist model. On the one hand, the state was being downsized and state capacity was being weakened. On the other hand, the private sector and the various (labor, capital, and finance) markets were expanding rapidly and the Chinese economy was reintegrating into the global capitalist economy.

Like other neoliberal states, China suffered considerable cost during her initial march toward neoliberal capitalism in the 1980s. A decade of market "reforms" already led to many serious economic problems, such as inflation, unemployment, corruption, and tax evasion. Inflation was over 30 percent in 1988 and 1989 when the state tried to decontrol commodity prices. Unemployment became a problem when bankrupted enterprises discharged workers. Workers showed signs of discontent as reforms began to exert tighter control over work schedule and raised work quotas. A government source estimated that 70 percent of the enterprises became rich through profiteering and speculation, while another source revealed that the private sector had evaded 70 to 80 percent of their taxes (So and Hua 1992).

In the late 1980s, the above economic problems and social grievances had triggered a democracy movement that led to a confrontation between the protesters and the party-state in the Tiananmen Square. The Tiananmen Incident was a first major challenge to the Chinese communist party-state during the post-Mao era. It led to bloody suppression of the protesters and serious political division within the party-state between the so-called reformist faction (which is pro-neoliberal reform) and the conservative faction (which is skeptical of such reform). What then happened after the Tiananmen Incident?

Re-building the State and the Deepening of Neoliberal Capitalism in the 1990s

In contrast to the image of a weakened state in the neoliberal literature, the Chinese state has considerably strengthened its managerial and fiscal capacity during the aftermath of the Tiananmen Incident. A new "cadre responsibility system" was instituted in the early 1990s by the central party-state to strengthen its control over the evaluation and monitoring of local leaders. County party secretaries and township heads sign performance contracts, pledge to attain certain targets laid down by higher levels, and are held personally responsible for attaining those targets. There are different contracts for different fields, such as industrial development, agricultural development, tax collection, family planning, and social order. The Chinese party-state has the capacity to be selective, that is, to implement its priority policies, to control the appointment of its key local leaders, and to target strategically important areas. Thus, Maria Edin (2003, 36) argues that "state capacity, defined here as the capacity to control and monitor lower-level agents, has increased in China, and that the Chinese Communist Party is capable of greater institutional adaptability than it is usually given credit for."

In addition, the state has strengthened its fiscal capacity. The central party-state introduced a "Tax Sharing Scheme" (TSS) in 1994 to redress the center-local imbalance in fiscal matters (Yep 2007). The TSS is aimed at improving the center's control over the economy by increasing "two ratios"—the share of budgetary revenue in GDP and the central share in total budgetary revenue. It seems that the TSS did succeed in raising the "two ratios" (Loo and Chow 2006), thus helping to arrest the decline of fiscal foundation of the center and increase the extractive capacity of the central party-state. Zheng (2004, 118–119) argues that the TSS has shifted fiscal power from the provinces to the center, so "now, it is the provinces that rely on the central government for revenue."

In addition, in contrast to the neoliberal doctrine's calling for less intervention, the Chinese state has intervened more in the economy. It has engaged in debt-financed investments in huge megaprojects to transform physical infrastructures. Astonishing rates of urbanization (no fewer than forty-two cities have expanded beyond the 1 million population mark since 1992) have required huge investments of fixed capital. New subway systems and highways are being built in major cities, and 8,500 miles of new railroad are proposed to link the interior to the economically dynamic coastal zone. China is also trying to build an interstate highway system more extensive than America's in just fifteen years, while practically every large city is building or has just completed a big new airport. These megaprojects have the potential to absorb surpluses of capital and labor for several years to come (Harvey 2005, 132). It is these massive debt-financing infrastructural and fixed-capital formation projects that make the Chinese state depart from the neoliberal orthodoxy and act like a Keynesian state.

Furthermore, after the party-state had strengthened its capacity and played a more active role in upgrading the economy, it also pushed for a deepening of neoliberalism. In the first wave of neoliberal reforms in the 1980s, the reform policies were aimed mostly to expand the private sector; they had left the public sector largely intact. Thus the reformers in the 1980s used the term "market socialism" to stress that China was still socialist because it had a dominant public sector and the party-state was still in control of the strategic sectors (or the commanding heights) of the Chinese economy.

However, the party-state turned to the public sector and pushed forward the following policies in the late 1990s:

- *Privatization and corporatization* policy to cut the size of the state sector and to increase the size of the private sector. In the 1990s the state-owned enterprises (SOEs) were undergoing corporatization, so they were no longer dependent on the state for funding, and they had to operate independently in the market. After corporatization, the SOEs were asked to run like an independent private profit-making enterprise; they can go bankrupt if they were losing money (So 2005). The SOEs were given the green light to lay off workers, to increase work intensity and productivity, and to cut workers' benefits if they found it necessary to remain competitive in the market. In the late 1990s, there observed the layoff of millions of state workers and the cutting back of their benefits.
- *Commodification of human services.* Whereas the Maoist state provided human services (like housing, health care, welfare, education, pension, etc.) based on need and free of charge to all citizens, the post-reform state treated human services as a commodity to be distributed to people on market principles. Housing, for example, is no longer provided to the state workers free. Instead, workers are now asked to find their own housing in the newly emerged private housing markets. Likewise, workers are now asked to pay a part of the costs for services in most welfare fields and social insurance, such as pension, medical care, and the newly created unemployment insurance, higher education, and many personal services (Guan 2000).
- *Deepening of liberalization.* Petras (2006) points out that China joining the World Trade Organization (WTO) is likely to lead to a further dismantling of the state sector, a dismantling of trade barriers and removal of subsidies, the savaging of the countryside, the near unquestioning orientation toward the export market strategy, and consolidation of foreign production as the leading force in the Chinese economy (see also Hart-Landsberg and Burkett 2004).

Social Resistance to the Deepening of Neoliberal Capitalism

As a result of the deepening of neoliberal policies, class inequalities expanded and class conflict rapidly intensified between labor and capital.

On the one hand, there is the formation of a cadre-capitalist class as a result of privatization/corporatization of state assets. Since the old Chinese capitalist class was eliminated in the 1950s, a new class of capitalist entrepreneurs had to be created in order to promote market reforms. During the first decade of the economic reforms (in the 1980s), in which a private sector was created, cadres (state officials) turned local state and collective enterprises into profitable Township and Village Enterprises (TVEs), developed joint ventures with foreign capitalists and overseas Chinese capitalists, quit their official positions to set up their own capitalist enterprises, and hired their kin and friends to run the new enterprises. Since cadres possessed political capital as well as the necessary networks to run their enterprises, they had an edge over other classes in taking advantage of the nascent business opportunities in the first decade of the reform era. It is this cadre-capitalist class that advocated the deepening of neoliberal policies in the aftermath of the 1989 Tiananmen Incident.

During the second stage of reform in the 1990s, when the state called for the privatization of state enterprises (through shareholdings)—with its "Grasp the big, release the small" policy—the assets and profits of state enterprises were diverted on a massive scale into the private hands of the cadres in charge of them. Ding's (2000) studies show that state enterprises were stripped in three ways: through organizational proliferation, consortium-building, and "one manager, two businesses." In organizational proliferation, cadres removed the best-equipped or most profitable segments of an enterprise and established collectively owned companies. Consortium-building refers to a partnership between economic entities in which a state-owned enterprise sets up a new firm in collaboration with a non-state-owned enterprise. "One manager, two businesses" happens when cadres establish their own private business by usurping assets from the state industries in which they continue to hold executive positions. Francis (2001) points out that these practices are carried out by all sorts of state entities—"local and municipal governments, national ministries, the army, national and local public security bureaus, party organizations, universities, scientific institutes." The coexistence and interpenetration of various forms of ownerships between the state and the non-state domain have provided a golden opportunity for cadres to transform themselves into capitalist owners and managers of semi-state, collective, and private properties.

As a result of the above practices, the emerging state-capitalist relationship is characterized by the fusion of the political capital of the cadres, the economic capital of the capitalists, and the social/network capital embedded in the local society. Many collective enterprises are owned and run by capitalists, while many private enterprises are spun-off state properties owned and run by state managers or their kin. This fusion makes it very hard to distinguish what is owned by the state, by the collective, or by the capitalist in the private sector, because the boundaries of their property relations are often blurred. The fuzzy property boundaries and the mutation between state managers and capitalists have created an all-powerful hybrid that can be called a "cadre-capitalist" class (So 2003).

At the same time, the deepening of neoliberal policy also has produced the formation of a working class. In the 1990s, the need to boost productivity and bring profits into the state sector led to attempts to lay off redundant workers, hire temporary and contractual workers from the rural migrants, cut wages, reduce workers' benefits, charge workers for services, intensify workloads, and enforce strict work discipline in order to improve the state enterprises' productivity and profitability. Workers in the state sector are now beginning to feel like proletarians in a capitalist enterprise.

In response to the above neoliberal policies, the Chinese working class has become restless. *China Labour Bulletin* (2002, 1) reported at the time that "almost every week in Hong Kong and mainland China, newspapers bring reports of some kind of labor action: a demonstration demanding pensions; a railway line being blocked by angry, unpaid workers; or collective legal action against illegal employer behavior such as body searches or forced overtime." According to the official statistics, in 1998 there were 6,767 collective actions (usually strikes or go-slows with a minimum of three people taking part) involving 251,268 people. This represented an increase in collective actions of 900 percent from the early 1990s. In 2000, this figure further jumped to 8,247 collective actions involving 259,445 workers (*China Labour Bulletin*, 2002, 2). Given such widespread labor protests, it is no wonder that the Chinese government identified the labor problem as the biggest threat to social and political stability (So 2007).

The peasants in the countryside too became restless because of increasing amount of tax and levy imposed by corrupt cadres in the local government. Thornton (2004, 87) cites a Chinese government report confirming that over 1.5 million cases of protest had occurred in 1993, over 6,000 of which were officially classified as "disturbances" (*naoshi*) by Chinese authorities. Of these cases, 830 involved more than one township and more than 500 participants; 78 involved more than one county and over 1,000 participants; and 21 were considered to be "extremely large-scale" events involving more than 5,000 participants. A surprising number of these confrontations turned violent: during these disturbances, 8,200 casualties resulted among township and county officials, 560 county-level offices were ransacked, and 385 public security personnel were fatally injured (So 2008).

Aside from the workers and peasants, there was also resistance from middle-class intellectuals. The late 1990s saw the emergence of many kinds of social movements (namely, the environmental movement, the consumer movement, the homeowners' resistance movement, the women's movement) in China (Economy 2005; Cai 2005; Chen 2003). Misra (2003) points to the rise of a group of critical intellectuals, the so-called "new left" (*xin zuopai*), who are highly dissatisfied with the growing socioeconomic class inequalities and the alarming decline of public morality. They show a greater appreciation for the Chinese revolution and wanted a reassessment of Western models of development (including modernization theories and neoliberalism).

Thus, when neoliberal reforms were deepened during the late 1990s, workers, peasants, and the middle class were getting restless, their criticisms of the problems of neoliberalism were more upfront and blunt, and their protests and demonstrations were becoming more widespread and violent. These societal responses were reflected in the party-state. In June 1998, thirty-five members of the elite Standing Committee of the National People's Congress (NPC) presented an emergency resolution to the top leadership of the Chinese Communist Party (CCP), accusing the government and Party of violating workers' "right of existence" and "trampling the worker-peasant alliance," and alluding to widespread protests and opposition to China's program of economic liberalization (Liew 2001; Nonini 2008).

The above challenges to the party-state happened at the right time because the party was then undergoing an elite transition. In 2002, President Jiang Zemin, who served as China's top leader for more than thirteen years, retired. Jiang was the one who proposed "The Three Representatives" policy to recruit more politically progressive people from the private sector into the communist party. Jiang's leadership team was replaced by Hu Jintao and Wen Jiabao. According to Joseph Cheng (2007), Hu and Wen's ideal is to return to the good old days of the 1950s when the Maoist Party was in full control, and the vast majority of Party cadres were uncorrupt, dedicated, and selfless.

By the early 2000s, Hu/Wen began to revise the neoliberal policies in response to all sorts of social resistance in Chinese society.

Backward Toward State Developmentalism

In contrast to the neoliberal doctrine that calls for the dismantling of the welfare state, the Chinese party-state under Hu/Wen leadership presented a new policy of "building a new socialist countryside" and a "harmonious society" in 2006 (Saich 2007). This policy is significant because it could signal a change in ideological orientation of the Chinese state (Kahn 2006). Whereas the pre-2006 Chinese party-state adopted a neoliberal orientation, it is now moving toward a more balanced one between economic growth and social development. While market reforms would continue, this new policy indicates that the state would play a more active role in moderating the negative impacts of marketization. In the new policy, the state will need to include "the people and environment" in its developmental plan, and not just focus narrowly on GNP indicators and economic growth.

Thus the new policy advocates a transfer of resources from the state to strengthen the fiscal foundation of the countryside. Not only was the agricultural tax abolished to help relieve the burden on farmers, but the state increased its rural expenditure by 15 percent (to $15 billion) to bankroll guaranteed minimum living allowances for farmers, and an 87 percent hike (to $4 billion) for the health care budget (Liu 2007). These policies indicate a massive infusion of funds from the state to the peasants and rural areas. In addition, there is a de-commodification of human services. Rural residents would no longer have to pay many miscellaneous charges levied by schools; fees at primary schools will be abolished as part of a nationwide campaign to eliminate them in the countryside for the first nine years of education. The state will also increase the subsidies for rural health cooperatives, which will be available in 80 percent of the rural counties by 2008. For now, rural residents have to pay market rates at the villages' private clinics, and most of them do not even have medical insurance and spend more than 80 percent of their cash on health care (Liu 2007). Furthermore, the new policy is aimed at reducing social inequality, especially the widening gap between the countryside and the city. Thus, pensions are to be made available for everyone, not just those enjoying a privileged status as registered urban residents. Over the past two years, the state has also been promoting the spread of Minimum Living Standard Assistance for the rural population. This is potentially a highly significant development, opening up for the first time the real possibility of instituting a social safety net that covers the whole of the population, whether urban or rural (*Economists*, 2006; Hussain 2005).

Like other developmental states in East Asia, China has a strong state machinery. Although a cadre-capitalist class has emerged at the local level when the state managers were asked to promote local development—so-called "local state corporatism" (Oi 1992)—this cadre-capitalist class has failed to capture the central party-state. Thus, the central party-state can still uphold the moral high ground of state socialism, going after the capitalists for tax evasion and the breaking of environmental laws, standing on the side of the workers by strengthening the labor laws, and standing on the side of peasants by cutting rural taxes and relocating more resources to the country-side. The party-state at the center blames corrupted officials for causing social unrests at the local level. The central Chinese state is highly autonomous in the sense that it is not "captured" by vested economic interests at the local level. The old generation of capitalists was largely destroyed in the Communist Revolution and later in the Cultural Revolution. The nascent capitalist class that has just emerged in the market reforms of the 1980s and 1990s is too weak and too dependent on the state to pose any challenge. In addition, the Chinese state has the capacity to carry out its developmental plans. Since it owns the banks and controls the financial sector, it has powerful policy tools at its disposal that makes the cooperation of indigenous business more likely: access to cheap credit, protection from external competition, and assisted access to export markets are all levers that the Chinese state can use to ensure business compliance with governmental goals. Since the Chinese corporations have a high debt/equity ratio, even the threat of withdrawal of state loans would be serious.

Second, like other developmental states in East Asia, the Chinese state has actively intervened in the economy. The state has become the engine powering capital accumulation. Aside from debt finance and infrastructure construction, the Chinese central state also develops plans for strategic development, decrees prices and regulates the movement of capital, and shares risks and underwrites research and development.

Third, like other developmental states in East Asia, the Chinese state has actively mobilized the ideology of nationalism and defines itself as carrying out a national project to make China strong and powerful. In the post-reform era, China was experiencing an ideological vacuum since the state could no longer be legitimized by Marxism or communism. Thus, nationalism became the state's only hope to get the support of the Chinese masses. The Chinese state seems to believe that the best response is to build a strong sense of national cohesiveness based on cultural heritage and tradition rather than to develop a nationalism based solely on hostility toward the outside world. Nationalism, however, can cut both ways. The Chinese state knows well that excessive nationalism might not only undercut the Communist Party's ability to rule but also disrupt China's paramount foreign policy objective of creating a long-term peaceful environment for its modernization program. The Chinese state's concern is reflected in its rejection of a more radical nationalism, such as that advocated by the authors of *The China that Can Say No*, as well as in its efforts to control anti-Japanese sentiment. Indeed, China's response to the provocation caused by Japanese leaders' visit of their controversial war shrine was far more restrained than it was in Taiwan and Hong Kong. The Chinese state's concern that nationalism had to be controlled was also evident in its efforts to restrain anti-Americanism in the aftermath of the NATO bombing of the Chinese embassy in Yugoslavia (Ogden 2003).

Fourth, like other developmental states in East Asia, the Chinese state adopts authoritarian policies to discipline labor, suppress labor protests, and to deactivate civil society in order to maintain a favorable environment to attract foreign investment and to facilitate capital accumulation. It seems authoritarianism is unavoidable in export-led industrialization because labor subordination is an important means to cheapen labor and to make the working class docile. Otherwise, the exports of the East Asian developmental states would not be competitive in the world economy, and transnational corporations would not relocate their labor-intensive production to East Asia. It is ironic that the Chinese state, with its tightly organized party-state machinery, has proven to be very effective in co-opting labor activists, dividing the working class, and silencing labor protests.

Finally, like other developmental states in East Asia, China received an influx of capital during its initial phase of capitalist industrialization. During the cold war era in the 1950s and 1960s, the massive influx of U.S. aid, loans, and contracts greatly helped East Asian states (South Korea, Taiwan) to solve the problem of initial accumulation, and it had greatly enhanced their states' capacity to promote developmental policies. The United States, of course, would not provide similar aid, loans, and contracts to China to assist its developmental program after the fading of the cold war in the 1980s and 1990s. Fortunately, there was a comparable influx of Chinese diaspora investment to China at the initial phase of transition to provide capital for initial accumulation. Before 1978, Chinese diaspora capitalism thrived in Hong Kong, Taiwan, Singapore, and other overseas Chinese communities. After the Chinese state adopted an open-door policy for foreign investment, Hong Kong accounted for the bulk of China's foreign investment and foreign trade. In the early 1990s, Hong Kong firms employed over 3 million workers in the Pearl River Delta. By the end of the 1980s, Taiwan became the second largest trading partner and investor for Mainland China. In the 1990s, overseas Chinese entrepreneurs in Southeast Asia have shown a visible interest in conducting trade and investment in China.

In short, China's latest developmental pattern is closer to that of the East Asian developmental state than to the neoliberal state. It has strong state machinery with a high degree of state autonomy and a strong capacity to carry out its goals. It greatly intervenes in the economy through developmental planning, deficit investment, export promotion, and strategic industrialization. It is also highly nationalistic and authoritarian, suppressing labor protests and limiting popular struggles. In addition, its capitalist industrialization has greatly benefited from an influx of capital during the critical phase of original capital accumulation.

Nevertheless, China's state developmentalism has also shown some significant differences from that of other East Asian states. First, the Chinese developmental state has exhibited a strong tendency toward entrepreneurship. Although East Asian state officials are promoting the hatching of capitalists, they seldom turn themselves into capitalists and involve in running the corporations. In China, however, not only were state officials asked to be good managers and turned state enterprises into profit-making businesses, but many state officials also informally turned public assets into quasi-public, quasi-private properties, or simply into private companies. As is well documented in the China field, there is a fuzzy boundary between state enterprises and collective/private enterprises, and it is difficult to draw a clear boundary between state officials and private capitalists in China. Rather, the Chinese characteristic is a hybrid "state-capitalist" walking on two legs in both the state sector and the private sector.

Second, the Chinese developmental state has exhibited a pattern of local, "bottom-up" strategy. East Asian developmental states had adopted a centralized policy, and it was their central governments that played the most active role in development. However, in China, due to the legacy of communism, the policy of fiscal decentralization, and the vast territory of China, local officials in provincial, county, and village governments have played a much more active role than their counterparts in East Asian developmental states. Instead of promoting the development of urban industrialization and mega cities, Chinese local state officials have promoted the development of rural industrialization and small and medium cities. In South China, for example, a new "bottom up" development mechanism is taking shape in which initiatives are made primarily by local states to solicit overseas Chinese and domestic capital, mobilize labor and land resources, and lead the local economy to enter the orbit of the international division of labor and global competition.

Third, although the Chinese developmental state has relied on economic growth and nationalism as its bases of legitimacy, it has also paid more attention to egalitarianism than its East Asian counterparts during their industrial takeoff. Having gone through the legacy of revolutionary socialism under the Maoist regime, and having a constitution that still claims that workers and peasants are the masters of society, the Chinese state was much more vulnerable to the charges of inequality, poverty, and exploitation than its East Asian counterparts. Thus, the Chinese state had many times backed off from carrying out policies that could lead to mass layoffs and the elimination of the social safety net. In its latest policy in 2006, the Chinese state aims to build a new socialist countryside, abolish agricultural tax, infuse funds to the peasants and the rural areas, and attempt some de-commodification policies that provide free education, subsidized health care, guaranteed minimum living standard, and instituting a safety net that covers the entire population.

Explaining the Transition from Neoliberalism to State Developmentalism

David Harvey (2005, 1) points out that 1978–1980 is a turning point in China's social and economic history. In 1978, Deng Xiaoping, the leader of the Chinese Communist Party, took the first momentous step toward the liberalization of a communist-ruled economy. The path that Deng defined was to transform China in two decades from a closed backwater to an open center of neoliberal capitalism in the global economy. The first decade of neoliberal reforms, however, had led to serious economic and social problems in Chinese society, triggering off the robust democracy protests at the Tiananmen Square in 1989 to challenge the rule of the Chinese Communist Party.

In the aftermath of the 1989 Tiananmen Incident, the party-state tried hard to restrengthen itself. It instituted a "cadre responsibility system" to improve local governance and a "tax sharing scheme" to readdress the center-local imbalance in fiscal matters. After political order and economic growth had been restored, the party-state determined to push for a deepening of neoliberal capitalism (such as privatization, commodification of social services, and the entry into WTO) in the mid-1990s.

By the late 1990s, however, China began to feel the pains of a neoliberal economy. First, there was super-exploitation of labor power, particularly of young women migrants from rural areas. Wage levels in China were extremely low, and conditions of labor, which was not sufficiently regulated, were despotic and exploitative. Moreover, China became one of the world's most unequal societies. Neoliberal market reforms had quickly transformed conditions in China into disparities in income among different classes, social strata, and regions, leading rapidly to social polarization. Formal measures of social inequalities, such as the Gini Coefficient, had confirmed that China had traveled the path from one of the most equalitarian societies to one with chronic inequality, all in the span of twenty years (Harvey 2005, 143). Furthermore, as usually happens in a country going through rapid capitalist industrialization, the failure to pay any attention to the environment was disastrous. In China, "rivers are highly polluted, water supplies are full of dangerous cancer-inducing chemicals, and public health provision is weak (as illustrated by the problems of SARS and the avian flu)" (Harvey 2005, 174). Edward Friedman (2007, 2) also points out that "China has a ruthless free market, no regulation, no safety standards, no FDA, no CDC, no NIH. It is also the world leader for people dying in industrial accidents, and about 400,000 each year die from drinking the water which is polluted."

In the late 1990s, the above contradictions had led to discontent and social conflict in society, as shown by the increasing call to regulate the market and by the growing numbers of labor protests, peasant demonstrations, social movements, and other large-scale social disturbances.

In the light of the above contradictions and discontents, the Chinese Communist party-state reconsidered its approach to neoliberal policies since the 1990s. In addition, neoliberalism was increasingly coming under attack and losing its creditability in the global economy. In the East, the "shock therapy"—which called for the dismantling of the centrally planned economy as soon as possible—not only did not work but also led to the downfall of the communist states in Eastern Europe. In the West, the antiglobalization movement was greatly empowered by its success in Seattle. In China, the Chinese party-state began in the late 1990s to reverse its neoliberal policies and started to build up a developmental state. After the party-state strengthened its fiscal capacity, it engaged in debt-financing investments in huge megaprojects to transform infrastructures and declared a new policy of "building a new socialist countryside" to address the issues of poverty and inequality in the rural areas.

However, the situation in China was not desperate. The Chinese state was not under any threat of foreign invasion, did not incur any large amount of foreign debt, and faced no immediate threat of any rebellion from below. As such, the Chinese state still had the autonomy and capacity to propose and implement various developmental policies "from above." For instance, the state could selectively introduce different types of developmental policies, could vary the speed of the market reforms, could expand or limit the space of opening up to transnational capital, and, most importantly, could still have the freedom of adjusting (or even reversing) its policies if they were not working.

The asymmetrical power relationship between the state and other classes has also given the state a free hand to try different developmental policies over the past few decades. The capitalist class was too small, too weak, and too dependent on the state to be the agent of historical transformation in China. The capitalist class is politically impotent to capture the state and carry out a neoliberal path of development. Facing growing labor unrest and popular struggle against such abuses as child labor in the coal mines, discrimination against immigrant workers, and environmental degradation, the capitalist class remains powerless to stop the policies toward state developmentalism.

Nevertheless, the transition from neoliberalism to state developmentalism took the form of a transition, not the form of a rupture or a revolution. The transition took a fairly long period of time and it was a gradual, adaptive process without a clear blueprint. The reforms have proceeded by trial and error, with frequent midcourse corrections and reversals of policies. In other words, Chinese state developmental policies were not a completed project settled in "one bang," but an ongoing process with many midcourse adjustments.

Situated in East Asia, China has long been attracted to the developmental state model that has achieved a remarkable postwar economic growth in South Korea, Taiwan, and Japan. Thus, Chang Kyung-Sup (2007) points out that there has been a conscious process of learning and transplanting technologies, industrial organization, and state policies among the East Asian states, and China is a leading example of this.

Future Trajectory

If the Chinese experience is characterized by trial and error, midcourse corrections, and reversals of policies, what is the future trajectory of state developmentalism in China? There are several scenarios: return to socialism, return to neoliberalism, move to imperialism, and the consolidation of state developmentalism.

First, the Left would be interested to know whether there is any possibility for China to return to socialism. Given the fact that China has moved away from socialism for almost thirty years and its capitalist-oriented economy has firmly institutionalized, it seems highly unlikely that socialism can make a dramatic come back in China. Besides, the Chinese working class and the peasants are still disorganized and are deprived of class organizations to protect their interests.

Second, another scenario is the return to neoliberalism. Harvey (2005) points out that neoliberalism is the project of the capitalist class through which it could exert its hegemony in advanced capitalist countries. Following this line of argument, the capitalist class will not be content to remain a junior partner of the developmental state forever. As soon as the capitalist class has matured and consolidated its power, it will push forward with its neoliberal project. In South Korea, for example, there was a dismantling of the Korean developmental state when the *chaebols* (big business corporations) were strengthened by their interlinkages with transnational corporations in the 1990s. This global reach has made the *chaebols* so powerful that they were able to dismantle the Economic Planning Board, set up private non-state financial institutions, and push for financial liberalization (Chiu and So 2006).

Although at present the Chinese capitalist class is still small and weak, it could grow very fast and become a force to challenge the party-state in a few decades. If this happens, the Chinese capitalist class will probably follow the path of its Korean counterpart: it will no longer be content to be a junior partner of the developmental state. Instead, it will expand its economic interests and push forward its neoliberal project.

The third scenario is the imperial path. State developmentalism becomes so successful that it greatly empowers China in the world economy. When China expands, it will inevitably run into conflict with other hegemonic states. When this occurs the great powers in the global economy will fight China over control of markets, resources (especially oil), technology, finance, and territory. History tells us that the existing hegemon will always want to hold onto its power and will try every means to prevent other states from challenging its position. Unless China can win this battle of hegemonic transition, it will not emerge as the center of capital accumulation in the twenty-first century. State developmentalism, by drawing upon national symbols and building up a strong state, does provide an impetus toward the above scenario of the rise of China and Chinese hegemonic struggles in the world economy.

Finally, there is the scenario of consolidation of state developmentalism. This paper argues that state developmentalism emerged in the late 1990s as a response to the developmental problems and social protests triggered by the deepening of neoliberal capitalism. In the late 2000s, China is again facing very serious developmental problems and social protests triggered by the global financial crisis. In November 2008, Roubini (2008) reported that China may be on its way to a hard landing, as the last batch of macro data from China all point toward a sharp deceleration of economic growth, sharply falling spending on consumer durables, falling home sales, and sharp fall in construction activity. Factories are closing in China's export region and unemployment is a growing concern in the urban areas. China needs a growth rate of at least 5 percent to absorb about 24 million people joining the labor force each year. The collapse of export trade has left millions without work and set off a wave of social instability. The *Sunday Times* reported on Feb 1, 2009 that social unrest among unemployed workers is spreading more widely in China than officially reported.

In response, the party-state quickly unveiled a US$586 billion stimulus plan (roughly 7 percent of its gross domestic product) over the next few years to improve infrastructure (to build new railways, subways, and airports) and to rebuild communities devastated by an earthquake in the Southwest in May 2008. The stimulus plan would cover ten areas, including low-income housing, electricity, water, rural infrastructure, and projects aimed at environmental protection and technological innovation—all of which could encourage consumer spending and bolster the economy. The party-state wants to promote domestic consumption and to improve collective consumption (such as expanding the health care network, lowering tuition fees for schools/universities, and upgrading the rudimentary social safety net) and social insurance. The assumption is that unless the social safety net and social insurance are expanded, the Chinese consumers will be more inclined to save than to spend, and the enlarged domestic market will not be able to absorb the slack in the export market caused by the global financial crisis of 2008–2009.

In addition, Hu Jintao, when giving a speech in December 2008, pointed out that "China should continue to hoist high the great flag of socialism with Chinese characteristics and push forward the significance of Marxism." In several CCP meetings in late 2008, President Hu also called on the armed forces and police to pull out the stops to uphold social stability by putting down disturbances and assorted conspiracies spearheaded by anti-China forces. Willy Lam (2009) labels the above policies as "The Great Leap Backward" because they are signaling a sharp U-turn from the neoliberal policies of the late 1990s.

In March 2009, Ching Chong reports that the party-state set up a special "6521 Group" (the numbers refer to the 60th anniversary of the founding of communist China, the 50th anniversary of the Tibetan uprising, the 20th anniversary of the Tiananmen crackdown, and 10th anniversary of the crackdown on the Falungong movement) and has issued a notice detailing thirty-three measures that governments at every level must take to protect public order when they are dealing with such economic threat as the "highly dangerous mob events" triggered by land grabs, massive urban and rural unemployment, labor disputes, and public discontent over the sale of fake or unsafe goods.

The thirty-three measures provide for a system of societal control to be implemented when the need arises. These measures allow authorities to conduct "orderly and effective control" over the Internet and online communities. Noteworthy too is how the official Xinhua news agency last month surrendered its power of deciding which foreign agency news could be transmitted in China. That authority went to the State Council Information Office. The activities of all non government organizations, whether domestic or foreign, as well as new social and economic organizations, are to be closely monitored (Ching 2009).

Facing sharp economic downturn and growing social unrest, the party-state has abundant reasons to move away from neoliberal capitalism to state developmentalism. Since the global economic crisis has just begun and China has just introduced a stimulus plan, it is obviously too early to tell whether the plan will work. However, if China does continue to move toward the path of state developmentalism, it could end up in a position that Silver and Arrighi (2000, 69) had envisioned a decade ago: "China appears to be emerging as the only poor country that has any chance in the foreseeable future of subverting the Western-dominated global hierarchy of wealth." Whether China succeeds in achieving this will depend on, to a large extent, how global capitalism is able to deal with the unfolding worldwide economic and financial crisis that is now threatening the very survival of the global capitalist system.

References

Burkett, Paul and Martin Hart-Landsberg. 2005. "Thinking about China: Capitalism, Socialism and Class Struggle." *Critical Asian Studies* 37, 3: 433–440.

Cai, Yongshun. 2005. "China's Moderate Middle Class: The Case of Homeowner's Resistance." *Asian Survey* 45, 5: 777–799.

Chan, Kam Wing. 2003. "Migration in China in the Reform Era: Characteristics, Consequences, and Implications," in Alvin Y. So., ed, *China's Developmental Miracle:Origins, Transformations, and Challenges*, 111–135. Armonk, NY: M. E. Sharpe.

Chang Kyung-Sup. 2007. "Developmental Statism in the Post-Socialist Context: China's Reform Politics through a Korean Perspective." Paper presented to the conference Chinese Society and China Studies at Nanjing, May 26–27, 2007.

Chen, An. 2003. "Rising Class Politics and Its Impact on China's Path to Democracy." *Democratization* 10, 2: 141–162.

Chen Xiangming. 2005. *As Border Bend: Transnational Spaces on the Pacific Rim*. Lanham, MD: Rowman & Littlefield.

Cheng, Joseph Y. S. 2007. "Introduction: Economic Growth and New Challenges," in Joseph Y. S. Cheng, ed., *Challenges and Policy Programmes of China's New Leadership*, 1–35. Hong Kong: City University of Hong Kong Press.

China Labour Bulletin. 2002. *Hong Kong: Han Dongfang* (January 31) *http://www.china-labour.org.hk/public/main*

Ching, Chong. 2009. "Govt aims to Minimise Fallout from key Political Anniversaries and Slump." *CHINAPOL Digest*, March 2, 2009. Special issue (114).

Chiu, Stephen and Alvin Y. So. 2006. "State-Market Realignment in Post Crises East Asia: From GNP Developmentalism to Welfare Developmentalism?" Paper presented to the conference on The Role of Government in Hong Kong at the Chinese University of Hong Kong, November 3.

Ding Xueliang. 2000. "Systemic Irregularity and Spontaneous Property Transformation in the Chinese Financial System." *China Quarterly* 163: 655–676.

Economist. 2006. "Asia: Dreaming of Harmony; China." *The Economist* 381, 8500. October 21.

Economy, Elizabeth. 2005. "China's Environmental Challenge." *Current History*, September: 278–283.

Edin, Maria. 2003. "State capacity and local agent control in China: CCP Cadre Management from a Township Perspective." *The China Quarterly* 173: 35–42.

Francis, Corinne Barbara. 2001. "Quasi-Public, Quasi-Private Trends in Emerging Market Economies." Comparative Politics 22: 275–294.

Friedman, Edward. 2007. "Living Without Freedom in China." *Footnotes: The Newsletter of Foreign Policy Research Institute* 12, 10, available on Web site: www.fpri.org. July 2007.

Halliday, Fred. 1976. "Marxist Analysis and Post-Revolutionary China." *New Left Review* 100: 165–192.

Hart-Landsberg, Martin and Paul Burkett. 2004. "China and Socialism: Market Reforms and Class Struggle." *Monthly Review* 56, 3: 1–116.

Harvey, David. 2005. *A Brief History of Neoliberalism*. New York: Oxford University Press.

Hussain, Athar. 2005. "Preparing China's Social Safety Net." *Current History* 104, 683: 683–722.

Kahn, Joseph. 2006. "A Sharp Debate Erupt in China Over Ideologies." *New York Times*, March 12: 1,1.

Lam Willy. 2009. "Hu Jintao's Great Leap Backward." *Far Eastern Economic Review*, January/February: 19–22.

Liew, L. H. 2001. "What is to be Done? WTO, Globalization and State-Labour Relations in China." *Australian Journal of Politics and History* 47, 1: 39–60.

Liu, Melinda. 2007. "Beijing's New Deal." *Newsweek*, March 26, 2007.

Loo Becky and Sin Yin Chow. 2006. "China's 1994 Tax Sharing Reforms: One System, Different Impact." *Asian Survey* 46: 215–237.

Misra, Kalpana. 2003. "Neo-Left and Neo-Right in Post Tiananmen China." *Asian Survey* 43, 5: 717–744.

Nonini, Donald M. 2008. "Is China becoming NeoLiberal?" *Critique of Anthropology* 28, 2: 145–176.

Ogden, Suzanne. 2003. "Chinese Nationalism: The Precedence of Community and Identity over Individual Rights," in Alvin Y. So, ed., *China's Developmental Miracle: Origins, Transformations, and Challenges*. Armonk, NY: M. E. Sharpe.

Oi, Jean. 1989. *State and Peasant in Contemporary China: The Political Economy of Village Government*. Berkeley: University of California Press.

———. 1992. "Fiscal Reform and the Economic Foundations of Local State Corporatism in China." *World Politics* 45: 99–126.

Petras, James. 1997. "The Cultural Revolution in Historical Perspective." *Journal of Contemporary Asia* 27, 4: 445–459.

———. 2006. "Past, Present and Future of China: From Semi-Colony to World Power?" *Journal of Contemporary Asia* 36, 4: 423–441.

Pun Ngai. 1999. "Becoming Dagongmei: The Politics of Identity and Difference in Reform China." *China Journal* 42: 1–19.

Ross, A. 2004. *Low Pay High Profile: The Global Push for Fair Labor.* New York: The New Press.

Roubini, Nouriel. 2008. "The Rising Risk of Hard Landing in China." *Japan Focus,* November 4.

Saich, Tony. 2007. "Focus on Social Development." *Asian Survey* 47: 32–43.

Silver, Beverly J. and Giovanni Arrighi. 2000. "Workers North and South," in Leo Panitch and Colin Leys, eds., *2001 Socialist Register: Working Classes, Global Realities.* New York: Monthly Review Press.

So, Alvin Y. 2003. "The Making of a Cadre-Capitalist Class in China," in Joseph Cheng, ed., *China's Challenges in the Twenty-First Century.* Hong Kong: City University of Hong Kong Press.

———. 2005. "Beyond the Logic of Capital and the Polarization Model: The State, Market Reforms, and the Plurality of Class Conflict in China." *Critical Asian Studies* 37, 3: 481–494.

———. 2007. "The State and Labor Insurgency in Post-Socialist China," in Joseph Cheng, ed., *Challenges and Policy Programs of China s New Leadership.* Hong Kong: City University of Hong Kong Press.

———. 2008. "Peasant Conflict and the Local Predatory State in the Chinese Countryside." *Journal of Peasant Studies* 34, 3–4: 560–581.

So, Alvin Y. and Shiping Hua. 1992. "Democracy as an Antisystemic movement in Taiwan, Hong Kong, and China: A World System Analysis." *Sociological Perspectives* 35, 2: 385–404.

So, Bennis Wai Yip. 2005. "Privatization," in Czeslaw Tubilewicz, ed., *Critical Issues in Contemporary China.* New York and London: Routledge.

Thornton, Patricia. 2004. "Comrades and Collectives in Arms: Tax Resistance, Evasion, and Avoidance Strategies in Post-Mao China," in Peter Hays Gries and Stanley Rosen, eds., *State and Society in 21st-Century China,* 87–104. New York and London: Routledge Curzon.

Wang Shaoguang and Hu Angang. 2001. *The Chinese Economy in Crisis: State Capacity and Tax Reform.* Armonk, NY: M. E. Sharpe.

Yep Rey. 2007. "Tax Assignment Reform and its Impact on County Finance: Enhancing Regulatory Capacity for the Central State?" Paper presented to the International Conference on "State Capacity of China in the 21st century," City University of Hong Kong, April 19–20, 2007.

Zheng Yongnian. 2004. *Globalization and State Transformation in China.* New York: Cambridge University Press.

8

Globalization and Gender: Women's Labor in the Global Economy

*Lourdes Benería**

Much has been written on the subject of gender and globalization. The emphasis in this chapter is on women's employment and on the global processes that have been affecting it. The rapid formation of a female labor force across the globe during the past decades has, to a great extent, been tied in particular to the growth of the service sector and of low-cost manufacturing, even though these have not been the only sectors behind the feminization of the labor force. The links between gender and globalization should not be seen as responding only to structural and economic forces; although these have, of course, been at the root of this feminization, they have also been shaped by the interaction between these forces and the different ways through which gender constructions have been used and reconstituted during the past decades. The feminist movement, in its quest for gender equality, has contributed to this trend on the supply side by emphasizing the need for women to search for greater financial autonomy, bargaining power, and control over their lives. But other tendencies have been at work, both on the supply and the demand side.

Trends in Women's Employment

Since the late 1970s, studies have documented a preference for women workers in different sectors, particularly in the service sector and in export-oriented, labor-intensive industries relying on low-cost production for global markets. Globalization has intensified these trends in

---

* This chapter has been compiled through excerpts from chapters 3 and 4 of the author's book *Gender, Development, and Globalization* (Routledge, 2003), selected and modified by the editor.

many areas. In its initial steps, the body of research that documented these trends tended to focus on the jobs created by transnational corporations in low-wage industrializing areas such as Southeast Asia. The emphasis was placed on the exploitation of women by transnational capital and on its ability to take advantage of female stereotypes associated with women workers: docility, nimble fingers, youth, often of rural origins from developing countries, acceptance of low wages, and poor working conditions. This analysis reflected a "women as victims approach," which gradually was seen as simplistic and unable to deal with the complexities involved (Lim 1983; Pyle 1982; Elson and Pearson 1989). Lim, for example, noted that women's employment in transnational corporations did result in improvements in their lives. Various authors began to point out the ways in which women were not passive victims of exploitative conditions and illustrated the multiplicity of factors that affected their incorporation in paid work and their active involvement in it (Ong 1987).

As a result, this initial period was gradually replaced by analyses of female employment that captured the complexities and the often contradictory effects involved (Elson and Pearson 1989). Studies since then have also focused on forms of female employment other than that provided by transnational capital—including its linkages with local capital through subcontracting and informal employment (Benería and Roldán 1987; Kabeer 2000). In contrast to the women-as-victims approach, the emphasis in many studies has been on illustrating the multiplicity of effects associated with women's participation in the labor force, including the gains resulting from women's increased autonomy and bargaining power as a result of employment. In Naila Kabeer's words, women's paid work has been associated with an increase in the "power to choose," even if within the many still existing constraints facing those she calls "weak winners" (Kabeer 2000). Likewise, it has resulted in women's ability to act and defend their interests and those of their family and community in the face of most adverse circumstances. This type of empirical work has taken place throughout a wide range of historical contexts, cultural practices, and gender constructions.

Women's Employment in the Export-Processing and Service Sectors

A significant proportion of studies of women's employment has continued to focus on low-wage production for export where female labor tends to concentrate. Such is the case with export-processing zones (EPZs) and informal employment in low-wage, labor-intensive manufacturing; the latter includes, for example, lower-tier subcontracting chains, micro-enterprises, and self-employment (Carr et al, 2000; Benería 2003, Ch. 4). Both rely on systems of flexible production that find in women's labor the most flexible supply, such as in the use of temporary contracts, part-time work, and unstable working conditions, which are at the heart of low-cost production for global markets and are tied to the volatility of global capital's mobility in search of the lowest cost location.

The service sector has also absorbed a large proportion of female employment, which can be subdivided in various categories:

- Expanding services associated with global markets tend to employ low-skill women in *pink-collar offices*, for example, for data entry and data processing in mail order business, airlines and rail systems, credit card providers, and other financial services like banking and insurance. These activities can be highly concentrated like in the case of the Caribbean and in some Asian countries such as China, India, Malaysia, and the Philippines. Referring to the case of Barbados, Carla Freeman has written about this offshore clerical work in the Caribbean as resulting in "a convergence between realms of tradition and modernity, gender and class—where transnational capital and production, the Barbadian state, and young Afro-Caribbean women together fashion a new 'classification' of woman worker who, gendered producer and consumer, is fully enmeshed in global and local, economic and cultural processes" (2000, 22). Women's employment has also expanded in the tourist sector across countries. Some estimates indicate that the proportion of women in these services is as high as in the export sector, and it is almost completely female in the case of the Caribbean. Needless to say, employment in this sector tends to be seasonal and unstable, depending also on the ups and downs of international demand.

- Globalization has also facilitated international networks linked to *prostitution and related services*. This is a sector for which reliable data is difficult to obtain. Nevertheless, existing estimates show that it has been growing in size and significance across the globe. In addition, the increasing phenomenon of child prostitution, male and female, has also become a matter of growing concern, and here, too, the numbers vary widely according to the source. An international debate has emerged around the extent to which sex workers chose this profession, and therefore should not be viewed as victims but in charge of their own circumstance and choices (Doezema and Kempadoo 1998; Outshoorn 2004). In any case, international prostitution raises difficult questions in terms of human rights and the means to prevent minors from being drawn into it. Sex tourism is one of the sectors where international migration and prostitution are linked. This is the case with the above Asian countries, but it involves other regions as well (Ehrenreich and Hochschild 2002). An analysis of the economic base of prostitution is crucial to understanding its different forms and manifestations through class-related labor market segmentation and working conditions. As Lim has pointed out, "policy makers have to deal with an industry that is highly organized and increasingly sophisticated and diversified, as well as having close linkages to the rest of the national and international economy" (1998, 9). Policy and action with regard to prostitution is for the most part addressed to the prostitutes (mostly female) rather than to their clients (mostly male), and to the institutions linked to the industry. In addition, policy and action should address the roots of poverty that feed the industry.

- Increasing migration by women from low- to high-income countries during the past decade has been getting much international attention. Much has been written about the large number of *domestic and daycare workers* from developing countries, supplying their labor to elite families or middle-class families with working mothers. Pushed by their search for a better life—imagined or real—migrant married and unmarried women from the Philippines, Sri Lanka, Mexico, Ecuador, Peru, and other Latin American countries have been working in the United States, Canada, and the European Community, as well as the Middle East, Hong Kong, and other high-income countries. The feminization of international migration has received increasing attention during the past few years and a growing number of studies are addressing its roots and consequences (Salazar Parreñas 2001; 2008; UNFPA 2006; Benería 2008). The crisis of care in high-income countries with a high participation of women in the labor market is at the root of this phenomenon, which we are only beginning to understand in terms of its long-term consequences. Migrant women, finding international employment more easily than the men in their communities, often have been leaving their family behind—including their own children. In her study of children from the Philippines left behind and cared by fathers, older siblings, and other family members, Salazar Parreñas has pointed out the negative consequences of the loss of maternal care and the changes that this generates in their lives. She concludes that, although migration often increases the standard of living of families, the children of migrant Filipina domestic workers suffer from the extraction of care from the global South to the global North, a pattern affecting many countries. On the other hand, migrant women contribute to meet the needs of Northern families in their efforts to reconcile family and labor market work. However, they leave a care vacuum behind, thus creating new care provision needs in their own countries (Benería 2008).

The feminization of the labor force has taken place even in countries where women's participation in paid work was traditionally low and socially unacceptable. The speed at which this phenomenon has occurred has raised interesting questions about the processes through which traditions and gender constructions can be dismantled or reconstituted and adapted to economic change. This has produced an interesting body of literature that analyzes the tensions and contradictions involved in the process (Pyle 1983; Ong 1987; Feldman 1992 and 2001; Kabeer 2000).[1] In this respect, sociological and cultural studies have made a rich contribution—incorporating levels of analysis that combine the more strictly economic aspects of globalization and women's employment with a focus on changes in gender relations, social constructions in the division of labor, women's agency, and household-market connections. Some authors, for example, have analyzed the phenomenon of prostitution within the framework of the different religions while others have suggested a close link between prostitution and the survival circuits facilitated through global cities (Lim 1998; Sassen 1998).

The Impact of Globalization on Women's Labor

Regarding the question of whether we can generalize about the gender effects of globalization, at least the following two points can be made. First, the literature has emphasized the notion that globalization and the feminization of the labor force have been parallel to the processes of labor market deregulation and flexibilization registered across countries during the past four decades. This has affected both men and women, although not necessarily in the same ways. Feminization has been parallel to the deterioration of working conditions and the race to the bottom resulting from global competition (Standing 1989 and 1999). Although some have interpreted this view as blaming this deterioration to women's new roles in production, its most common interpretation emphasizes the key role of women's cheap labor to deal with the pressures of international competition and global markets. At the same time, although by far the largest proportion of women's jobs are located at the lower echelons of the labor hierarchy, the observed economic polarization among women, together with North-South differences, reflect the fact that a small proportion of women has gained a relatively advantageous position in the global economy. Thus, class and other differences need to be taken into consideration as well.

Second, generalizations about the effects of globalization on women must be approached with great caution since effects vary according to historical, socioeconomic, and other conditions. To illustrate, the variety of studies that over the years have analyzed the effects of export-oriented manufacturing in Southeast Asia since the 1970s have shown that the high level of female employment generated has in the long run resulted in improvements, even if far from spectacular, in women's earnings and a higher degree of gender equality (Lim 1983; Dollar and Gatti 1999; Seguino 2000). Yet, the Asian experience cannot be applied to other countries. For instance, the maquiladora sector in the U.S.-Mexican border represents a model of export-oriented production that over the years has not resulted in gains for the large majority of women employed (Cravey 1998; Fussell 2000). Fussell's study for the case of Tijuana, Mexico, found that, in their drive to keep production costs low, transnational manufacturers have tapped into women's low-wage labor, "thereby taking advantage of women's labor market disadvantages and making a labor force willing to accept more 'flexible' terms of employment" (2000, 59). Differences between these outcomes are due to varying factors having to do with labor availability (relatively limited in the case of the Asian countries and practically unlimited in the Mexican case), degrees of wage inequality, and the locally specific dynamics of the labor market with respect to male/female employment.[2]

As a result, a debate has been generated on the relationship between export-oriented growth and women's wages, on the one hand, and working conditions and gender equality, on the other. Those who hold a more optimistic view of the connections between the two have argued that gender inequality has been reduced in terms of wage differentials, access to jobs, and educational achievement (Dollar and Gatti 1999). On the other hand, those who take a less optimistic view argue that, for example, in the case of the Asian tigers, economic growth was correlated to wage gender gaps, that is, growth was fed by gender inequality. Taking the second position, Seguino (2000) has shown that the Asian economies that grew most rapidly had the widest wage gaps. Similarly, Hsiung (1995) illustrates how Taiwan's high level of flexibility and market adaptability was solidly based on low wages and poor working conditions of women as home-based workers.

The Informal Economy

During the 1970s and early 1980s, what had been called initially the informal "sector" was viewed in developing circles as a transitory form of employment. It was conceptualized as "backward" and in contrast with the formal sector, which was viewed as the "modern"

solution to the low productivity and poor working conditions prevalent at the informal level. It was also assumed that, as countries developed, the formal sector would absorb most informal activities and the
marginal working population (ILO 1972; SSP/UCECA 1976). These
initial formulations emphasized the informal sector's connections
with the marginality of the urban poor as well as their unstable working conditions and their precarious location within the economy. The
SSP/UCECA study defined it in reference to the following factors:

- Low skills and productivity
- Very low level of earnings
- Absence of, or very precarious, job contracts
- Unstable working conditions
- Poor access to social services and absence of fringe benefits
- Very low rates of affiliation to labor organizations
- Illegal or quasi-legal work

Over the course of over three decades, we have witnessed an increasing
reliance of firms and households on precarious forms of employment
and a deterioration of labor market conditions for a large proportion
of the workforce. Far from diminishing its importance, the informal
"sector" has become increasingly larger in size and its composition
more complex, to such an extent that the ILO began calling it "informal economy" in the early 2000s. To be sure, we need to distinguish
between two types of informalized activities: those linked directly
or indirectly to industrial and service work in more formal settings
and those representing survival activities organized at the household
and community level. The former are linked to profit-oriented operations and can include self-employment and wage work tied directly or
indirectly to more formal production processes. This sector includes
micro-enterprises and subcontracting arrangements, both in high-
and low-income countries. Survival activities, on the other hand, tend
to represent the most precarious forms of self-employment with weak
or no links to the more formal processes and without possibilities for
any degree of capital accumulation. These are, in fact, the most visible
activities in the urban landscapes of developing countries' cities.

Another type of differentiation between these sectors results
from the legal/illegal divide. The informalization of labor processes
observed in high-income countries as a result of globalization has, to
a great extent, taken place within the context of legality, with important exceptions located in the underground economy. The many temporary agencies that deal with contingent work operate within the

confines of legality, even if they might not offer the protections of full-time employment. This is much less the case in developing countries where, despite its growing importance, the informal economy lacks legal status and work takes place under the usual precarious conditions that have traditionally been associated with the sector. As a result, workers engaged in informal activities tend to have little or no access to social protection and other benefits. Neither are they covered by national labor legislation. Hernando De Soto has argued that these are the activities that are "filling the vacuum left by the legal economy" (2000, 49) because of its market rigidities and controls. Far from being absorbed by the formal economy, as had been assumed in the 1970s, informal activities have been on the increase during the past three decades. In Latin America, for example, where labor markets have been deeply transformed since the 1980s, most observers agree in the diminishing centrality of formal employment. As Pérez-Sáinz (1999; 2000) has pointed out, there are a variety of reasons for its diminishing relative importance in the region, from the effects of structural adjustment and market deregulation to the weakening of public employment due to budget cuts and privatization programs. Thus, although Latin America traditionally had high levels of informal employment, the past three decades have registered further growth; in urban areas, it represented 47.9 percent of total urban employment in 1998, up from 44.4 percent in 1990 (ILO 1999). A recent Inter-American Development Bank (IDB) study (2004) indicated that, in the mid-1990s, the micro-enterprise sector employed more than 50 percent of the labor force in most Latin American countries and that, between 1990 and 1995, an average of 84 out of 100 new jobs in the region were generated by micro-enterprises. By 2009, more than half of total urban employment in Latin America was accounted for by the informal economy, and micro-enterprises employed the great majority of people engaged in work in this sector.

These trends have led to the growing reliance on precarious forms of survival across countries, particularly for the poorest households but also affecting other sectors (Oliveira 2000; Gonzalez de la Rocha 2006). Household survival strategies include very unstable links with the labor market, combining wage labor and self-employment—often within short time periods—as well as temporary migration (domestic and international). This instability led Bolivian sociologist Garcia-Linera (1999) to talk very appropriately about the phenomenon of "nomad labor," referring to survival strategies based on moving from one job to another or from one location to another. International migration, for example, in the case of Bolivian workers working temporarily in Argentina and Brazil, had become an important part in these often precarious strategies up until the crisis in Argentina. In international development circles, the literature has used the notion of "labor exclusion" to refer to the vicious circle of poverty resulting from persistent levels of unemployment, underemployment, and marginality from regular sources of income. For example, weighted averages for Latin America showed a 9 percent rate of open urban unemployment for 1999, "a figure above the 8.3 percent for 1985, at the height of the debt crisis" (Pérez-Sáinz 2000).[3]

Informalization and Women

Women comprise a significant proportion of workers engaged in informal activities, although this varies from region to region. Hampered either by low marketable skills to their credit or by other obstacles, such as lack of mobility and the need to combine work with child care and domestic activities, many women from poor households go into informal activities to generate whatever income they possibly can. At least since the 1970s, women have been highly involved in informal activities, but given the labor market transformations of the past four decades, we can ask whether there have been changes in the extent and nature of this involvement. Several observations can be made in this regard.

First, the feminization of the labor force during the past three decades has intensified the reliance of many women on informalized employment. Self-employment reflects this trend and the proportion of self-employed in the female nonagricultural labor force increased in all regions, including the "developed regions" (Charmes 2000). Although statistical information regarding the scope of informal activities where women concentrate is deficient, studies have shown that they range from subcontracting processes linked to export-oriented industrialization—including home-based work—to street vending and other trade and service activities that evolve around survival strategies.

Subcontracting and home-based work illustrate many of the problems associated with women's informal employment. A study of subcontracted work in five Asian countries (The Philippines, Thailand, India, Pakistan, and Sri Lanka) shows that earnings lower than in the formal sector prevail, with no consistency in work contracts, difficult working conditions, and long hours of work (Balakrishnan and Huang 2000). The study points out the difficulty of organizing workers for the purpose of increasing their bargaining power, and it illustrates that "subcontracting makes it very difficult to hold one employer responsible for protecting workers' rights" due to "the many layers of chains" (Balakrishnan and Huang 2000, 14). Several studies mention that married women with children are often preferred by subcontracting firms (Dangler 1994; Boris and Prugl 1996). Due to their limited mobility and narrower range of options in the labor market, married women in particular offer greater labor force security for firms (Hsiung 1995). Although they cannot have direct control over work done at home and, hence, they cannot directly monitor workers, firms can take advantage of the discipline imposed on women by their need to both remain at home to care for children and other domestic activities, and to earn whatever income they can.

Another study differentiates between two types of home-based workers—"independent own-account producers" and "dependent subcontract workers"—pointing out that the term "homeworkers" refers to the second category (Carr, Chen, and Tate 2000). Women represent the large majority of home-based workers in many areas, reaching beyond the 80 percent level in some countries. Although the variations between countries are large, women represent a significant proportion of the nonagricultural labor force. Home-based work in the service sector has also been expanding in high-income countries such as France, the United Kingdom, and the United States, where much of this work is of clerical nature, including typing, word processing, editing, and telemarketing (Carr, Chen, and Tate 2000).

Second, in the North as well as the South, at least three aspects of economic restructuring have implications for the informalization of work, with a variety of gender dimensions:

- At the micro level, industrial restructuring has profoundly transformed the linkages between core firms and the different levels at which production has been decentralized. Production through subcontracting provides examples across countries and industries. A study of the shoe industry in Spain's Mediterranean region, for example, has illustrated the ways in which the larger firms have reduced their size through the formation of smaller firms and through decentralized production based on more informal labor contracts, many of which have gone underground. Ybarra (2000) has estimated that total employment has been halved since the 1980s, despite the increase in the number of firms. Women have concentrated at the lower levels of production, particularly in home-based work in which labor norms are "rarely implemented"

(Ybarra 2000, 213). The underground work carried out by women has been estimated to comprise between 35 and 40 percent of the work generated by the sector.

- Layoffs and relocation of production do not necessarily affect male and female workers in the same way. As a case study of the Smith-Corona relocation from Cortland, New York, illustrates, transitions after layoffs can result in gender-related differences in income loss, length of unemployment, transitional strategies adopted, and in the impacts from layoffs and unemployment experienced at the household and community level (Benería 2003, Ch. 4).
- The literature on "commodity chains" and peripheral urban growth has also contributed examples of the ways in which labor market informalization can affect women. Gereffi and others have analyzed the connections between globalization and the formation of commodity chains through which large buyers tend to control the links between inputs and outputs. (Gereffi and Korzeniewicz 1994). Along the same lines, Carr, Chen, and Tate (2000) pointed out that technological change has facilitated "lean retailing" that demands the "quick and timely supply of goods associated with the just in time inventory system" (2000, 126). According to these authors, this system has resulted in an increase in homework in the garment sector, particularly in countries close to the main markets of Europe and North America. Thus, the traditional precarious conditions in the informal sector have been reinforced by the dynamics of globalization in these new productive processes.

Third, women's primary involvement in domestic work and child care responsibilities continues to be a source of economic vulnerability for them, not only because this is unpaid work but also because it diminishes women's mobility and autonomy to design their labor market strategies. The effort of the past three decades to account for and analyze unpaid work and its consequences for women's participation in paid production has not been sufficiently translated into practical action and policies. In developing countries, middle- and upper-class households can rely on poor women to take up the responsibilities of domestic work and child care and to facilitate professional women's incorporation in market work. Domestic service still represents a very large proportion of women's informal employment in many low- and middle-income countries.[4] In the North, the crisis of care is often met with the hiring of immigrant women from the South, thus creating another care deficit for the families left behind (Ehrenreich and Hochschild 2002; Benería 2008).

In any case, involvement in unpaid work and child care responsibilities often ties women to informal employment and continues to have an impact on their choices and ability to participate in paid production on an equal basis with men—even if differences exist according to class and social background. We know that the implementation of austerity policies has tended to increase women's work and multiple responsibilities. However, these policies have not included appropriate provisions addressing the different problems faced mostly by women as a result of these responsibilities, nor have they taken into consideration existing legislation. A typical example of the latter is provided by the existence of ILO Convention 103 on maternity protection that has been ratified only by thirty-seven countries. This low rate of ratification points to the low priority given to this issue in most countries, let alone the fact that "maternity protection," in contrast to "parental protection," represents an intrinsic bias in the assignment of child care to mothers without equal share with fathers. With the increasing informalization of jobs, the implementation of ILO standards seems even more remote. In fact, within the framework of deteriorating labor conditions created by globalization, the ILO standards have been disregarded and subject to new scrutiny (ILO 1997).

Fourth, one of the differences between the earlier periods of informalized labor and the present time is the degree to which women have been able to take up actions at the national and international level. Structural adjustment and economic crises have led women to organize around labor issues as well as around tensions related to unpaid work and household survival strategies. An interesting example has been women's key role in getting the ILO Convention on Home Work approved in June of 1996. Elizabeth Prugl (1999) has argued that its approval was a feminist victory, with some international networks, such as HomeNet and the Self-Employed Women's Association (SEWA), providing extensive information and using the special relationship that its members had built with unions in advance of the conference to get their arguments on the floor. Although the convention's rate of ratification at the country level is very low, it provides concrete goals and a regulatory tool around which to organize further action. Organizations focusing on women and informalized work, such as SEWA, have gained international recognition for its accomplishments on behalf of homeworkers, particularly in India. Its increasing international influence has been important in the formation of other groups such as South Africa's Self-Employed Women's Union (SEWU). Similarly, Women and Informal Employment Globalizing and Organizing (WIEGO) has organized a network of workers, activists, and academics focusing on the informal sector, including home-based workers, and linking their interests and actions with research and work on improving statistical information on this sector at the global level.

Contradictory Tendencies for Women

During the past three decades there have been positive and negative changes for women that need to be taken into consideration in order to evaluate the complex and often contradictory tendencies affecting women's work. The effects of globalization and reorganization of the work process are highly uneven among women within and across countries. Thus, generalizations need to be qualified, particularly as the conditions affecting women's employment are very varied and often complex. There are several reasons behind this complexity.

First, gender gaps in education have been decreasing significantly across regions. For example, the Arab countries have experienced some of the most dramatic increases in women's educational indicators, with women's literacy rates doubling between 1970 and 1990. Southeast Asia and the Pacific countries also made very significant progress during this period. In many Latin American countries educational indicators for women have surpassed those of men (UNDP 1999).

There is much agreement on the notion that the improvement in women's educational status is a crucial step toward gender equality, women's advancement, and social development as a whole. However, while a correlation exists between schooling and labor force participation, and while this correlation tends to be higher for women than for men, women's educational achievements do not necessarily translate fully into labor market gains. Obstacles to women's advancement, such as those resulting from occupational segregation and gender-based discriminatory practices, reduce these possible gains. In addition, the progress made in women's education is far from complete. For example, gender differences in illiteracy rates and other indicators of educational achievement are still substantial in many countries. Illiteracy rates are extremely high in some African and Asian countries, while female primary and secondary school enrollment has not achieved parity with men in many areas.[5] To the extent that a high concentration of women in informalized production can be partially a result of their lower educational status, educational policies are crucial to deal with women's economic vulnerability and other aspects of women's lives. At a bare minimum, the elimination of illiteracy is an urgent objective for educational policies in the countries affected.

Second, there are clear indications that women's higher educational levels and rising labor market participation have contributed to a gradual, even if insufficient, increase in women's participation in managerial and professional occupations. The improved working conditions and social mobility of women in higher education stands in contrast with the precariousness and low-income levels received by the majority. This polarization seems to be at the root of growing income inequalities among women. Although more studies are needed to document this tendency for different countries, available evidence for Brazil and the United States points in the direction of what McCrate has called a "growing class divide among women" (Lavinas 1996; McCrate 1995). The more favorable real earnings for women and the decline in the pay gender gap in the United States have resulted in significant differences among them. As with men, the less educated are gradually falling behind as wage and earnings disparities by education have grown (Bertola et al. 2001, 267).

Third, despite the persistence of gender discrimination and obstacles to women's advancement, women's relative wages have tended to improve in relation to male wages across countries. According to UNIFEM (2000), during the period 1980 to 1997, in industry and services this improvement took place in twenty-two out of twenty-nine countries, and the list includes both high- and low-income countries. Similarly, for the same period in manufacturing industries, gender wage disparities decreased for twenty out of twenty-two countries. For the transition economies of Eastern Europe, an improvement in women's relative wages was registered for four out of seven countries. To be sure, this narrowing of the wage gender gap might be due to the fact that the relative wage for male workers has deteriorated. To add complexity to this issue, a UN (1999) report on the role of women in development pointed out that there is mixed evidence on whether the gender wage gap has increased or decreased. The report argues that in some countries, like the United States, the gap indeed seems to have narrowed, while in others, such as Japan, it appears to have widened. Similarly, the survey reports that trends vary among developing countries, with a narrowing of the gap in countries such as El Salvador and Sri Lanka and a widening in some Asian countries, such as Hong Kong, Singapore, and Taiwan.

Fourth, I have argued that much has changed since Ester Boserup (1970) emphasized the need to "integrate women in development." As she saw it, women had lost out in the process of development for a variety of reasons. One of them had to do with the ways in which industrialization, particularly under import substitution policies, had resulted in the marginalization of women due to the replacement of craft production with modern industry employing predominantly male labor. During the past decades, we have seen that the new preference for women workers, particularly in the manufacturing and service sectors, has contributed to the feminization of the labor force, thus reversing the trend mentioned by Boserup. Although the new processes of industrialization have provided many illustrations of the precariousness of women's employment, they have also contributed to raising women's income and autonomy, thereby generating contradictory results. Along these lines, we can distinguish at least between three different outcomes associated with the feminization of the labor force.

First are the cases that have generated gains for women. As mentioned above, under rapid growth and absorption of labor into labor-intensive export manufacturing, women have experienced wage increases as their share of industry employment has expanded. Contrary to the initial literature on the subject in the Southeast Asian region, Lim (1983) argued, with respect to the Southeast Asian countries, that rapid export-led growth benefited women by providing them with formal, well-paid employment. In particular, she argued that transnational firms paid higher wages than national capital. However, subsequent research has focused on the mass economic implications of this wage inequality, namely, that rapid growth in Southeast Asia was partly based on a high degree of gender and wage inequality. This is the case with Seguino's argument that low female wages served as an incentive for investment and exports, "by lowering unit labor costs, [and] providing the foreign exchange to purchase capital and intermediate goods which raise productivity and growth rates" (Seguino 2000, 29). Thus, the improvement in women's employment conditions in the Southeast Asian region needs to be evaluated within the wider framework of gender and wage inequalities.

Wage gains cannot be the only factor to evaluate the benefits of change for women. As Seguino has pointed out, women's socialization in Asia has led them to "accept their economic and social status, reassuring investors that labor strife will be unlikely," adding that "Women's lower wages and, in some cases, dismissal from employment upon marriage, have maintained their lower bargaining power not only relative to employers but also to men" (2000, 51). Hence, an evaluation of the effects of employment for women needs to take into consideration not only economic effects but also what happens at the level of gender socialization and power relations. Studies focusing on

sociocultural aspects of women's involvement in paid work have analyzed the different and often contradictory aspects of women's participation in the new processes of industrialization (Ong 1987; Feldman
1992). Their analyses of women's agency, of changing gender identities, and of women's capacity to contest oppressive practices—at work
or in their lives in general—have added important dimensions for an
evaluation of women's employment.

Second, cases where growth in female share of industry employment has been held in check and women's wage gains are limited.
An example is provided by Fussell's study, on the maquiladora industries along the U.S.-Mexico border. Using a Labor Trajectory Survey
for Tijuana, Fussell (2000) argues that maquiladora wages did not
improve over the years as employment expanded in the area. This
contradicts Lim's assertion that export-oriented employment raises
wages for workers and improves women's labor market position. As
already suggested, the difference between the two cases can be attributed to the conditions prevailing in Mexico in comparison with those
in Southeast Asia. While the maquiladora area continued to attract
an almost unlimited labor supply, the rapid growth in Southeast Asia
resulted in tight labor markets and high productivity increases, both
of which contributed to raising real wages. Three more factors have
contributed to this situation in the maquiladora industry: (1) economic
restructuring and the introduction of high-tech production systems
have benefited male labor, thus decreasing the proportion of women
workers from its peak in 1985; (2) due to high levels of unemployment and migration from rural areas in Mexico, young male workers
in particular have replaced women, particularly because they have
been willing to take up previously female jobs (Cravey 1998); and (3)
reports about the effects of China's membership in the WTO indicate
that countries like Mexico are suffering from the erosion of their relative comparative advantage as the much lower labor cost in the giant
Chinese economy threaten to shift production away.

An interesting study of the effects of export-led growth on gender
wage inequality in Taiwan provides another example (Berik 2000).
Using industry-level panel data, the study shows that economic
restructuring and technological change since the 1980s, together with
greater export orientation, shifted employment opportunities in manufacturing from wage to salaried employment. This was accompanied
by a disproportionate loss of employment opportunities for women
and an increase in gender wage inequality. Technological change
brought higher real wages for men and lower real wages for women.

Thus, in industries that underwent faster technological change, women wage workers experienced both absolute losses and losses relative to men. This reversal of feminization due to the introduction of new technologies—in the maquiladora industries, in Taiwan, and other cases—occurs for a variety of gender-biased reasons; they include men's greater opportunities to upgrade their skills and to benefit from the introduction of high technology—for example, through after-work training programs that are less accessible to women due to their domestic responsibilities. Likewise, women's difficulties in adjusting to schedules of flexible production are due to these responsibilities. Hsiung (1995) provides a similar evaluation of the Taiwanese experience from an ethnographic perspective.

Third, there are also cases of mixed results with increases in the female share of industrial employment, but under highly volatile conditions or contingent on the continuation of favorable circumstances for international capital. A study of gender differences in employment in Turkey's export-led industrialization provides an interesting example (Ozler 2001). Based on a large plant-level data set of Turkish manufacturing, Ozler argues that trade liberalization has led to the feminization of its labor force, with job creation for women significantly higher than for their male counterparts. However, the volatility of women's jobs is also significantly higher. Thus, this case reflects a preference for women workers, but for employment in jobs that are insecure. Mixed results are represented also by the process of defeminization described for the cases of Taiwan and the maquiladora industry, and similar changes have also been observed elsewhere.[6]

Finally, the pressures of global competition can create ambivalent situations with respect to women's employment. For example, Kabeer (2002) has argued strongly that the imposition of international labor standards can be detrimental for women workers in Bangladesh, on the grounds that implementation of higher standards might drive investment away from the country. Hence, the ability for women to benefit from international investment might be limited by the threat of adopting even minimum core standards.

Taken together, these cases imply that the initial 1970s literature emphasizing the adverse effects of industrialization on women's employment by global capital was simplistic at best. Global capital was, and continues to be, exploitative and causes disruptions and adverse effects in many cases, but its specific effects need to be examined case by case so as to take into consideration the range of variations in labor market conditions as well as the ways in which gender inequality affects the outcomes.

These examples also show that much has changed in terms of the profound transformations in gender roles in the workplace and outside of it, both for women and men. From a different perspective, Richard Anker's 1996 study of gender segregation across countries illustrated how, over a period of about three decades, men have been losing their labor market advantage—in the sense of having their "own" occupations protected against female competition. There are, however, some exceptions, and results differ between industries, countries, and regions. Anker's analysis for the OECD countries illustrates the extent to which the increase in women's labor force participation has taken place in female-dominated occupations, particularly in the 1970s, as well as in male-dominated occupations in the 1980s. We have come to view gender and gender differences in a dynamic way, reflecting its changing meanings over time. As illustrated by Matthew Gutmann in his ethnographic study of changing relations in Mexico City, "Gender identities, roles and relations do not remain frozen in place, either for individuals or for groups" (Gutman 1996, 27).This implies discarding stereotypes about the gender division of labor, employment conditions, and other factors affecting gender relations and gender differences. It also implies that a focus on only women is incomplete and often inaccurate for any type of gender analysis since it leaves out the changing nature of gender roles and locations, and their implications for power relations.

Notes

1. Some exceptions to the more general trends can be found in the economies of the former Soviet Union where the post-1989 period created contradictory tendencies. Women in these countries had registered very high labor force participation rates during the Soviet era, but they have suffered disproportionately from the social costs of the transition, including unemployment, gender discrimination, and reinforcement of patriarchal forms. In many cases, the transition to more privatized market economies reduced women's employment opportunities and relegated women to temporary and low-pay jobs (Moghadam 1993; Bridger et al. 1996; World Bank 2000). At the same time, the new market forces have generated jobs for women as a source of cheap labor, particularly in labor-intensive production for global markets. Hence, contradictory tendencies have been observed.
2. In the Mexican case, the proportion of women in the maquiladora labor force, which originally reached levels above 60 percent, began to decrease since the mid-1980s. This was due to several reasons, including technological shifts in

production toward more flexible production systems requiring new skills and increasing employment and availability of male labor (due to unemployment and migration, particularly of young males willing to work for low wages).

3. For some countries, unemployment rates in 1999 reached much higher levels, such as in Argentina (14.5 percent), Colombia (19.8 percent), Panama (13 percent), and Venezuela (15.3 percent) (Perez-Sainz 2000).

4. To illustrate with the case of Brazil, estimates of the proportion of employed women in domestic service range between 16 percent and 20 percent; one study found an average of 19 percent for the 1990s (Benería and Rosenberg 1999).

5. To illustrate, the 1997 female illiteracy rate was 97.1 percent in Ethiopia, 92.8 percent in Niger, and 79.3 percent in Nepal (UNDP 1999).

6. See, for example, an extensive study by Martha Roldan on the gender effects of economic restructuring in the auto industry in Argentina, illustrating the multiple links and meanings of the process of defeminization (Roldan 1994 and 2000).

References

Balakrishnan, Radhika, and M. Huang. 2000. "Flexible Workers—Hidden Employers: Gender and Subcontracting in the Global Economy, Report on a Research Project of the Women's Economic and Legal Rights Program." Washington, D.C.: The Asia Foundation.

Benería, Lourdes. 2003. *Gender, Development, and Globalization*. New York: Routledge.

———. 2008. "The Crisis of Care, International Migration, and Public Policy." *Feminist Economics* 14, 3: 1–21.

Benería, Lourdes, and Martha Roldán. 1987. *The Crossroads of Class & Gender: Industrial Homework, Subcontracting, and Household Dynamics in Mexico City*. Chicago: University of Chicago Press.

Benería, Lourdes, and F Rosenberg. 1999. "Brazil Gender Review." Report/evaluation of World Bank projects in Brazil.

Berik, Günseli. 2000. "Mature Export-Led Growth and Gender Wage Inequality." *Feminist Economics* 6, 3: 1–26.

Boris, Eileen, and Elizabeth Prugl (eds.). 1996. *Homeworkers in Global Perspective: Invisible No More*. New York: Routledge.

Boserup, Ester. 1970. *Women's Role in Economic Development*. New York: St. Martin's Press.

Bridger, Sue, Rebecca Kay, and Kathryn Pinnick. 1996. *No More Heroines? Russia, Women and the Market*. London: Routledge.

Carr, Marilyn, Martha Chen, and Jane Tate. 2000. "Globalization and Home-based Workers." *Feminist Economics* 3, 3, November: 123–142.

Cravey, Altha J. 1998. *Women and Work in Mexico's Maquiladoras*. Lanham, MD: Rowman and Littlefield.

Dangler, Jamie. 1994. *Hidden in the Home: The Role of Waged Homework in the Modern World Economy*. Albany, NY: State University of New York Press.

De Soto, Hernando, 2000. *El Misterio del Capital*. Lima: Empresa Editoria El Comercio.

Dicken, P. 1998. *Global Shift: Transforming the World Economy*. New York: The Guildfod Press.

Doezema, J., and K. Kempadoo (eds.). 1998. *Global Sex Workers: Rights, Resistance and Redefinition*. New York: Routledge.

Dollar, David, and Roberta Gatti. 1999. "Gender Inequality, Income, and Growth: Are Good Times Good for Women?" Washington, D.C.: The World Bank, Policy Research Group on Gender and Development, Working Paper Series, No. 1.

Ehrenreich, Barbara, and Arlie Russell Hochschild (eds.). 2002. *Global Woman: Nannies, Maids, and Sex Workers in the New Economy*. New York: Metropolitan Books.

Elson, Diane, and Ruth Pearson (eds.). 1989. *Women's Employment and Multinationals in Europe*. London: Macmillan Press.

Feldman, Shelley. 1992. "Crisis, Islam, and Gender in Bangladesh: The Social Construction of a Female Labor Force," in Lourdes Benería and Shelley Feldman, eds., *Unequal Burden: Economic Crises, Persistent Poverty, and Women's Work*. Boulder, CO: Westview Press, 105–130.

Fussel, M. E. 2000. "Making Labor Flexible: The Recomposition of Tijuana's Macquiladora Female Labor Force." *Feminist Economics* 6, 3: 59–80.

Gonzalez de la Rocha, M. 2006. "Private Adjustments: Household Responses to the Erosion of Work." New York: UNDP/SEPED Conference Paper Series.

Gutmann, Matthew. 1996. *The Meanings of Macho: Being a Man in Mexico City*. Berkeley: University of California Press.

Hsiung, Ping-Chun. 1995. *Living Rooms as Factories: Class, Gender, and the Satellite Factory System in Taiwan*. Philadelphia: Temple University Press.

ILO (International Labor Office). 1972. *Employment, Incomes, and Equality: Strategy for Increasing Productive Employment in Kenya*. Geneva: ILO.

———. 1997. *The ILO, Standard Setting, and Globalization, Report of the Director General*, 85th Session, Geneva: ILO.

Kabeer, Naila. 2002. *The Power to Choose: Bangladeshi Garment Workers in London and Dhaka*. London and New York: Verso.

Lavinas, Lena. 1996. "As mulheres no universo da pobreza: o caso Brasileiro." *Estudoes Feministas* 4, 2: 464–479.

Lim, Lin (ed.). 1998. *The Sex Sector. The Economic and Social Basis of Prostitution in South East Asia*. Geneva: ILO.

Lim, Linda. 1983. "Capitalism, Imperialism, and Patriarchy: The Dilemma of Third World Women Workers in Multinational Factories," in J. Nash and M. Fernandez-Kelly, eds., *Women, Men, and the International Division of Labor*. Albany, NY: State University of New York Press.

Moghadam, Valentine. 1993. *Democratic Reform and the Position of Women in Transitional Economies*. Oxford: Clarendon Press.

Oliveira, O., de. 2000. "Households and Families in a Context of Crisis, Adjustment and Economic Restructuring." Colegio de Mexico, Center for Sociological Studies.

Ong, Aiwa. 1987. *Spirits of Resistance and Capitalist Discipline: Women Factory Workers in Malaysia*. Albany, NY: SUNY Press.

Outshoorn, J. (ed.). 2004. *The Politics of Prostitution. Women's Movements, Democratic States and the Globalization of Sex Commerce*, Cambridge: Cambridge University Press.

Ozler, S. 2001. "Export-led Industrialization and Gender Differences in Job Creation and Destruction: Micro Evidence from the Turkish Manufacturing Sector." Unpublished paper, Economics Department, University of California at Los Angeles.

Pérez-Sáinz, Juan Pablo. 1999. *From the Finca to the Maquila: Labor and Capitalist Development in Central America*. Boulder, CO: Westview Press.

———. 2000. "Labor Market Transformations in Latin America During the 90s: Some Analytical Remarks." FLACSO: Costa Rica.

Prugl, Elizabeth. 1999. *The Global Construction of Gender: Home-Based Work in the Political Economy of the 20th Century*. New York: Columbia University Press.

Pyle, Jean. 1983. "Export-led Development and the Underemployment of Women: The Impact of Discriminatory Employment Policy in the Republic of Ireland," in J. Nask and M. P. Fernández-Kelly, eds. *Women, Men and the International Division of Labor*. Albany, NY: SUNY Press.

Roldán, Martha. 1994. "Flexible Specialization, Technology and Employment in Argentina: Critical Just-in-time Restructuring in a Cluster Context." Working paper WEP 2—22!WP. 240. Geneva: 1W.

———. 2000. *Globalización o Mundialización Teorla y Prdctica de Procesos Productivos y Asimetrias de Género*. Buenos Aires: FLACSO.

Salazar Parreñas, Rhacel. 2001. *Servants of Globalization: Women, Migration, and Domestic Work*. Stanford: Stanford University Press.

———. 2008. *The Force of Domesticity: Filipina Migrants and Globalization*. New York: NYU Press.

Sassen, Saskia, 1998. *Globalization and its Discontents*. New York: The New Press.

Seguino, Stephanie. 2000. "Accounting for Gender in Asian Economic Growth: Adding Gender to the Equation." *Feminist Economics* 6, 3, November: 27–58.

Seguino, Stephanie, Thomas Stevens, and Mark Lutz. 1996. "Gender and Cooperative Behavior: Economic Man Rides Alone." *Feminist Economics* 2, 1, Spring: 195–223.

SSP/UCECA (Secretaría de Programación y Presupuesto/Unidad Coordinadora del Empleo, Capacitación y Adiestramiento). 1976. *La Ocupación Informal en Areas Urbanas*. Mexico D.F.

Standing, Guy. 1989. "Global Feminization Through Flexible Labor." *World Development* 17, 7: 1,077–1,095.

———. 1999. *Global Labour Flexibility: Seeking Distributive Justice*. New York: St. Martin's Press.

UNDP (United Nations Development Programme). *Human Development Report*. Various years. New York: Oxford University Press.

UNFPA (United Nations Population Fund), 2006. *The State of the World Population; Women and International Migration.* New York: UNFPA.

World Bank. 2000. *World Development Report, 2000–01: Attacking Poverty.* New York: Oxford University Press.

Ybarra, J. A. 2000. "La informalozación como estrategia productiva. Un analisis del calzado valenciano," *Revista de Estudios Regionales* 57: 199–217.

9

The Failure of Neoliberal Globalization and the End of Empire

Martin Orr

This chapter offers a critical analysis of the development and dissolution of neoliberalism and neoconservatism, with emphasis on the role of popular resistance in bringing about the collapse of both forms of imperial domination. Responding to the brutal realities of the post-cold war "new world order," opposition to neoliberal globalization grew over the 1990s, culminating in the protests against the 1999 meetings of the World Trade Organization in Seattle. Having successfully blocked the machinations of the leading capitalist powers for world domination, Seattle served as a model for a series of worldwide protests and demonstrations that, despite state repression and media obfuscation, prevented the expansion of the neoliberal globalization agenda. At the same time, the increasing inability of oil production to keep up with the growth of emerging industrial economies has further intensified inter-imperialist rivalry. Under the cover of the attacks of September 11th, the neoconservative forces in the United States adopted a policy of overt imperial intervention in an attempt to gain control over its rivals' access to oil and thus secure its global domination. Although this turn at first deflated the antiglobalization movement, growing condemnation of the U.S. invasion and occupation of Iraq gave the anti-imperialist movement new life, and broadened its agenda. With vast majorities at home and abroad in opposition to U.S. aggression in Iraq and around the world, the failure of neoconservatism marks the end of empire.

In many ways, the 1999 protests during the Seattle meetings of the World Trade Organization (WTO) marked the failure in the United States of neoliberal, multilateral globalization and the ascendance of neoconservative, unilateral imperialism. Among economic and political elites, support for the agenda of the WTO and its sister institutions—the World Bank and the International Monetary Fund (IMF)—and for other pacts like the North American Free Trade Agreement (NAFTA) was in no way diminished. However, more bellicose foreign policies, toward allies and adversaries alike, contributed to a situation in which the United States became so isolated that it could no longer dominate the development of international trade agreements. Over a few short years, the U.S. government abandoned diplomacy as the vehicle by which corporate domination might be normalized, and in its place adopted a state of perpetual war of aggression. Instead of an era of globalization, we lived once again in an era of unabashed imperialism, which was perhaps not seen in the United States since World War I. In response, despite, and in part because of, a level of ideological hegemony and state repression unimaginable to many only a decade ago, resistance to U.S. domination intensified. The Battle of Seattle and the consequent failure of the neoliberal agenda stymied the push toward globalization, and then the international outrage over the U.S. invasion and failed occupation of Iraq placed into stark relief the limits of U.S. hegemony.

This chapter provides a critical analysis of the emergence and failure of neoliberal globalization in the face of international protest, the subsequent emergence and collapse of the neoconservative project, and the role of repression and resistance in these developments. The triumph of globalization in the 1990s is explained as the culmination of the *pax Americana* following World War II, a high water mark of U.S. international dominance made possible by the dissolution of the Soviet Union. For a relatively short time, under the veil of democratic consensus and the promise of a "global village" in which all humanity would share in the benefits of the "information age," the United States presented itself as the beneficent lone superpower, finally able to lead the world to peace and prosperity. The failure of this promise quickly became apparent, and a global movement of activists, nongovernmental organizations, and national, state, and local governments emerged and grew into an effective obstacle to neoliberal globalization. At the same time, a new threat to international trade emerged—an immanent peak in the rate of production of petrochemicals, especially oil, just as demand began to explode with the economic growth of China and India. The failure of neoliberal globalization and the impending global energy crisis required that the United States, if it was to retain its position of global dominance, would have to embrace a more aggressive foreign policy.

However, this policy failed in every respect. The rejection of neoconservatism and the optimism aroused internationally by the election of Barack Obama—itself possible only because of ground broken by progressive activism—will undoubtedly help create new opportunities. But every early indication suggests that there will be an attempt to return to pre-Seattle neoliberalism while offering generous concessions to the Bush-era neoconservatives. But, as became a talking point during the 2008 election, "the fundamentals of the economy are not sound." It is doubtful that the post-World War II era of U.S. domination will be restored.

Globalization, the Antiglobalization Movement, and the Failure of Neoliberalism

The institutions most closely associated with neoliberal globalization—the WTO, the IMF, and the World Bank—have their origins in the negotiations between business elites in the United States, the United Kingdom, and Germany in the 1930s (Leibovitz and Finkel 1997; Simpson 1993; Seldes 1943). Having agreed not to let war interfere with business, it didn't. U.S. corporations, including Chase Manhattan Bank, Ford Motor Company, Standard Oil, and IBM continued to do business with Nazi Germany (Higham 1983; Parenti 1997). Writing during the war, George Seldes was blunt: "[The] great owners and rulers of America...planned world domination through political and military Fascism, just as surely as Hitler did in Germany, and like groups and like leaders did in other countries" (Seldes 1943, 69). Little changed over the second half of the century.

With the settlement of World War II, the United States usurped from the European powers domination of their colonial empires, and used that position to place Europe itself in a relationship of dependency. A series of agreements were reached and institutions founded, all of which were dominated to a greater or lesser extent by the United States—of these the United Nations (UN) and the North Atlantic Treaty Organization (NATO) remain central. But as significant in the history of contemporary globalization were the Bretton Woods agreements. Bretton Woods established the World Bank and the IMF. At

Bretton Woods, an International Trade Organization (ITO) was also envisioned. Concerned that cold war competition for the allegiance of people and governments around the world would leave the United States unable to prevent the inclusion of provisions that would defend human, civil, and labor rights, the ITO was stillborn. Instead a weaker framework was established—the General Agreement on Tariffs and Trade (GATT). GATT became the forum through which further trade negotiations would take place.

The dissolution of the Soviet Union, the restoration of capitalism in the Warsaw Pact nations, and the unchallenged military superiority of the United States allowed the rebirth of the dream of an ITO. Unchecked by the Soviet alternative and with domestic dissent quiescent, the United States pushed for the creation of a suitably authoritarian body to enforce overarching global trade agreements. As the culmination of the Uruguay Round of negations of GATT, the WTO was established in January 1995. As of January 2007, there were 150 member nations, and another thirty-one had observer status (WTO 2007).

Administering existing trade agreements, arbitrating trade disputes, and providing a forum for continuing negotiations, the WTO promotes an international regime in which governments are discouraged from impeding "free trade" through tariffs, bans on imports and exports, and other "trade restrictive" laws that protect workers, consumers, and the environment. Minimum wage, prohibition of child labor, equal pay provisions, government set-asides, limitations on the length of the work week, workplace safety provisions, consumer safety, product labeling, environmental protection, and public services (including education and health care) are not excluded from being challenged as a violation of "free trade." An agenda this all-encompassing generated opposition from every quarter.

The demonstrations in Seattle were obviously not the first time that neoliberalism had been challenged. Still, the protests in Seattle were remarkable in that it mobilized an extremely broad, international, and highly visible coalition of forces. Labor, students, environmentalists, religious communities, and small business and farming interests exhibited a degree of solidarity rare in contemporary American history.

On November 30, 1999—"N30" as it was dubbed by organizers—about 40,000 demonstrators took to the streets of Seattle. Early that morning, protesters blocked intersections and off-ramps leading to the Convention Center. Riot police attempted to clear the streets using batons, pepper spray, tear gas, and rubber bullets. Later, there was a standoff in which hundreds of protestors occupied the intersection of 4th and Pike Streets. Although that afternoon a handful of demonstrators vandalized corporate targets like Niketown, Starbucks, and the Gap, they never threatened or perpetrated violence against delegates or law enforcement. As night fell, riot police pushed the crowd back, adding concussion grenades to their arsenal. A seven o'clock, curfew was imposed (Sunde 1999), and the following day Seattle Mayor Paul Schell suspended the rights to speech and assembly in downtown Seattle. Police were reinforced and hundreds of National Guard troops were deployed. Over 500 protesters were arrested for violating Schell's decree.

The Battle in Seattle had a number of consequences. First, and most immediately, many delegates were unable to reach the Convention Center, and the opening meetings of the WTO were delayed. Third World governments, many already opposed to some or all of the Millennial Round proposals (Weissman 1999), were emboldened by the support for their positions in the streets (and a year after the Seattle protests, the 133 members of the G77 group of developing nations expressed support for the protesters at the IMF meetings in Prague). The protest was wildly successful in that the Seattle talks failed.

Second, Seattle gave momentum to the movement against the WTO, its sister institutions, other trade conferences, and the political economic order driving neoliberalism. Seattle was reenacted over the following years in protests against the IMF and the World Bank in Washington, D.C. and Prague, and against the Miami meeting of delegates considering a Central American Free Trade Agreement. Since the North Atlantic powers continue to dominate international relations, demonstrations are now routinely staged against the G-8. In addition, the bipartisan support for "free trade" led to protests against the conventions of both the Republican and Democratic Parties in the summer of 2000, at the presidential debates, and at the inauguration of George W. Bush. Ralph Nader's candidacy in that year's Presidential election was in large part a response to the neoliberal agenda. Protests were also held against the mouthpieces of power at the National Association of Broadcasters meetings in San Francisco and Seattle.

Third, the response to the protest by the state and by media served to radicalize opposition, strengthen coalitions, and further broaden their demands. Experiencing firsthand the escalating violence of the state, and comparing the media whitewash of this violence with their own experience, led to the development of more sophisticated analyses of power, and of more effective strategies and tactics.

State violence, before and since Seattle, is never an accident. Preparations were made long before N30. In the weeks before, the FBI issued a security alert, and the Federal Emergency Management Agency (FEMA) issued its own, under the heading "terrorism" (Morales 2000, 63). Since Seattle, state repression of opposition has become increasingly preemptive and violent. Beyond batons and chemicals, tactics typically employed by the U.S. and European governments have included surveillance and infiltration, raids of offices and staging areas, preemptive arrests and detention of movement organizers, seizure of literature, mass arrests of nonconfrontational protesters, police brutality in custody, and live ammunition (Allen 2000). Observers immediately compared the new security regime to the FBI's COINTELPRO and the municipal red squads of decades past (Scher 2001). By the time of the Prague demonstrations against the IMF and the World Bank held on September 26, 2000, coordination among governments had become international, with lists of activists provided by the FBI to the Czech government to be used to prevent Americans from entering the country (Scher 2001). For a meeting of the industrial powers to take place, there must be a massive expenditure of resources, a militarized perimeter around the site, and the commission of political crimes.

The U.S. media had largely ignored the WTO prior to Seattle. Since then, the media have served to belittle and vilify those who oppose neoliberalism, and to cover up and facilitate state repression. Over the course of the past three decades, through mergers, acquisitions, and the deregulation of ownership rules, media have become increasingly corporatized and otherwise beholden to private interests (Bagdikian 2004; Compaine and Gomery 2000; Herman and Chomsky 1988; McChesney 2008; Parenti 1986). Corporate monopolization of media entails a unity of interests between media and telecommunication owners, such as General Electric, Disney/ABC, AOL/Time Warner, and News Corporation, who directly benefit from neoliberalism. In addition, these media giants sell audiences to corporate advertisers, who also benefit from the efforts of the WTO.

The increasing corporate concentration of media and their advertisers explains the crimes of omission and commission with regard to the WTO and other institutions of power, and with regard to the progressive coalitions challenging it. Sustained coverage of the costs of global inequality is effectively self-censored—the mythical benefits of globalization for working people are emphasized. On the other hand, the very tangible benefits that accrue to corporations—and the child labor, sweatshops, environmental fallout, and human rights violations that come as a part and parcel with globalization—are ignored (Baker 2000).

A centerpiece of this propaganda is the alleged "inevitability" of globalization (Ackerman 2000). Since one might as well oppose the rising of the sun, activists must be unrealistically utopian and uninformed. Prone to irrational violence, they are legitimate targets of police violence, and beneath contempt (Ackerman 2000; Coen 2000; Coen 2000a). Even on the rare occasions when police violence has been reported by the corporate press, it has been treated as either deserved or a result of mistaken identity (e.g., Trofimov and Johnson 2001). The extent of the surveillance and infiltration, the use of agent provocateurs, and the deliberate targeting of nonviolent protestors, legal observers, and journalists are ignored.

Experience with state and ideological repression created a more organizationally diverse and international movement, one that employs a wide range of often innovative and effective tactics. The movement has most closely been associated with the rallies and pickets, the guerrilla street theater, and the direct disruption of public and private meetings. But antiglobalization activists have also run third-party campaigns at all levels of government. They have made legal challenges, especially to police and prison misconduct, including the successful suits against the City of Seattle (McDonald 2007). Key to their success, activists nourished alternative and more participatory media outlets. Central among these is the Indymedia network (Flanders 2000). The Independent Media Centers (IMCs) were inspired in part by the Zapatistas' pioneering use of the Internet to disseminate information, organize activists, and attract allies in their struggle against the Mexican government and NAFTA. Indymedia was an important part of the Seattle effort, allowing people all over the world to bypass the corporate media, and to create and distribute their own news accounts, updates, commentary, and audiovisual files. Over a few years, the network grew to over one hundred autonomous collectives, each required only to ensure open publishing and a democratic process. It became the source for breaking news during a demonstration, and built a sense of community among an important core of activists. A host of new forms of participatory journalism and commentary have emerged alongside the more established progressive alternatives to the corporate media. With a broad coalition, sophisticated analyses of power, innovative uses of the media, and growing support from the public, the movement was able to build momentum, and could not be ignored.

In what one observer called the "Seattle effect" (Haski 2000, 9), the meetings of the IMF in Prague the following year "turned into a contest over which fat cat could show the most contrition for capitalism's shortcomings and evince the greatest sympathy for the poor" (Zachary 2000). There has been no substantive change, but this shift in rhetoric attests to effects of opposition. This attempt to make common cause with those fighting inequality has helped legitimize the movement and attracted additional support. No major meeting of political leaders and trade delegates has since gone unmolested, no meaningful progress in creating overarching U.S.-dominated trade agreements has been made, and the WTO abandoned the "Millennial Round" negotiations. But as trade talks stalled, and opposition grew, a new threat to business as usual was emerging.

Globalization and the Politics of Oil

Over the early history of the antiglobalization movement, through the Seattle protests, and certainly among most American activists, dialogue was largely confined to the role of the institutions established at Bretton Woods—the IMF and World Bank—and to the precursor of the WTO, GATT. Although most of the core organizers of this and other actions were quick to connect globalization and neocolonialism—and some were indeed willing to use the word "empire"— capital flight, child labor, environmental degradation, dismissal of consumer safety, and other human rights violations were the arguments that carried the day among more casual observers among the public. The suicide hijackings of September 11th, and the subsequent Bush Doctrine and the "preemptive" war against Iraq, thrust a new word into the international lexicon of dissent—oil.

Petrochemicals are the feedstock of the industrial age. The industrial revolution was powered by coal, and by the early twentieth century oil began to replace coal. Over the course of the century, the use of oil grew exponentially, to the point that it is now impossible for the denizen of any industrial society to get through the day without consuming profligate quantities of oil (Yeoman 2004, xi–xiii). When the first oil wells started producing in the late 1850s, the world population was a little over a billion. It is currently at nearly seven billion, and the laws of physics dictate that this global reduction in entropy has been made possible by cheap energy. As an energy source, and in petroleum-based plastics, pharmaceuticals, fertilizers, and pesticides, oil has changed the way we move people and goods, the way we grow and preserve food, the way we treat and distribute water, the way we build homes and furniture, the way we treat infection and perform surgery, and the way we conduct war. It has also shaped our political economy. In the United States, from J. D. Rockefeller to Halliburton—oil has been at the heart of wealth, and of political power.

However, access to oil—and the wealth and power that accrue to those who control its production and distribution—has never been assured. For nations that lack petroleum reserves sufficient to meet domestic demand, competition with other nations over reliable oil supplies has been one of the driving forces of their foreign policies. As the dependence of industrial society upon oil grew, access to oil has been explicitly described as a "vital national interest"—whether by Britain, Germany, or the United States.

The major wars of the past century have been fought over oil. In the late nineteenth century, the Baku deposits on the shores of the Caspian Sea ignited the first scramble for oil in Europe, and the British government's goal of preventing Germany from gaining access to oil through the Berlin to Baghdad railway better explains the cause of World War I than the assassination of an obscure nobleman (Engdahl 1993; Yeomans 2004). In 1912, amidst the naval arms race between England and Germany, Winston Churchill, then First Lord of the Admiralty, ordered the construction of new battleships to be fueled by oil rather than coal. As the threat of war loomed, the British government bought a majority stake in the Anglo-Persian Oil Company, now British Petroleum, in part to fuel the new oil-driven fleet (Klare 2001, 30). As war erupted, Britain's first act was to mobilize its forces in Basra. During the war, the United States supplied 80 percent of the Allies' petroleum, and "[as] Lord Curzon, a member of Britain's War Cabinet, triumphantly announced: 'The Allied cause had floated to victory on a wave of oil'" (Yeomans 2004, 9–10).

World War II was no less a war for oil. In Europe, Hitler's decision to divert military resources toward Baku contributed to the Soviet victory at Stalingrad, and Germany's retreat became a rout due to a shortage of oil to fuel its tanks and vehicles. In the Pacific, the United States' blockade of Japan, depriving it of steel and oil, precipitated the attack on Pearl Harbor (Klare 2001, 31). Pushing the Japanese from oil deposits in Dutch Indonesia toward the home islands and destroying its tankers cut off the oil essential for an industrial war economy.

At the conclusion of the war, the United States was in the driver's seat—literally. Unscathed by direct attack against home industry, and having had the ability to control access to the one essential resource of industrial warfare, the United States found itself in possession of half of the world's manufacturing capacity and held 70 percent of the world's gold reserves—all as a result of the fact that it produced nearly two-thirds of the world's oil. With all that oil, and backed by the gold and war debt that had siphoned in, the U.S. dollar became the international currency standard. At the time, there couldn't have seemed much choice among those making these decisions—since most purchases of oil and manufactured goods would be from the United States, it would be hard to do otherwise.

As financial hegemony passed from Britain to the United States, so too did the attempt to dominate the Middle East. In 1933, Saudi Arabia granted oil concessions to Standard Oil of California, and in 1943 the U.S. government began providing military aid to the Saudis. In February 1945, with the end of World War II in sight, President Franklin Delano Roosevelt met with King Abdul Aziz ibn Saud aboard the USS *Quincy* on station in the waters of the Suez Canal. The details of this meeting were never made public, but their effect was clearly to link U.S. protection of the Saudi regime with access to Saudi oil. With President Truman's 1948 letter to King ibn Saud, there was no mistake about the arrangement: "No threat to your Kingdom could occur that would not be a matter of immediate concern to the United States" (Yeomans 2004, 16). The Eisenhower Doctrine, framed as motivated only by a selfless interest in protecting "free nations" from Soviet communism, extended this commitment to the whole of the Middle East. Sufficient energy resources to meet domestic demand, and the countervailing power of the Soviet Union, moderated against direct military intervention in the Middle East through the 1960s (unflinching support for brutal regimes in Israel and Saudi Arabia and CIA machinations in Iraq and Iran notwithstanding).

Fostering, defending, and attempting to perpetually increase the rate of production of Saudi oil, the United States position as the world's filling station began to erode almost immediately. Since petrochemicals are finite resources, and given that only that which exists may be extracted, the ability of the United States to meet rising world demand could never be permanent. In 1970, with the United States already importing a third of the oil it consumed, the annual rate of

U.S. production of crude oil peaked. The following year, the Nixon Administration notified the world that the United States would abrogate the central tenet of the Bretton Woods agreements—no longer would U.S. dollars be redeemable for gold. In the following years, world governments abandoned the gold standard altogether and currencies were allowed to float based on market confidence in each nation's economy. At the same time, the Organization of Petroleum Exporting Countries (OPEC) emerged as a counterforce to Anglo-American oil corporations in global energy markets. American support of Israel's occupation of Palestine turned sentiment in the Middle East against the United States. With OPEC, and especially Saudi Arabia, now in the driver's seat, the U.S. economy began its downward spiral through the "stagflation" of the Carter Administration, and the recessions and "trickle down" austerity programs under the Reagan and Bush administrations, only to face the "jobless recovery" and, ultimately, the failure of neoliberalism under Clinton.

The inability to meet growing demand led to a demand-destroying bubble in oil and natural gas prices. Despite the respite offered by the subsequent global recession, production near capacity means that even periodic shortages due to mishaps, to environmental and climatological disaster, and to political and military conflict threaten to cause contemporary urban and industrial systems to collapse. A growing body of literature suggests that the rate of global extraction of petroleum is soon to peak—if it has not done so already (Campbell 2004; Darley 2004; Deffeyes 2001; Heinberg 2003; 2004; Klare 2001; 2009; Kunstler 2005; Simmons 2005). Most oil exporting nations are extracting less oil every year. Long the world's "swing producer," the ability of Saudi Arabia to ramp up production to meet shortfalls elsewhere is in question—between 2005 and 2006 its rate of production fell by 2.3 percent (British Petroleum 2007, 8). Saudi production in 2007 was down from 2006 by an additional 4.1 percent (British Petroleum 2008, 8).

Although oil corporations have justified price-gouging in terms of scarce supplies, and although the U.S. government and media have encouraged consumption as usual, those who most directly defend consumption exhibit an awareness of the ramifications of peak oil:

> The days of inexpensive, convenient, abundant energy sources are quickly drawing to a close....After peak production, supply no longer meets demand, and prices and competition increase....We must act now to develop the technology and infrastructure necessary to

transition to other energy sources and energy efficient technologies.
(U.S. Army Corp of Engineers 2005, 4)

This statement from a center of U.S. power should not be treated as
the most sophisticated analysis of the problem, nor should it be taken
at face value. Still, it is remarkable on a number of levels. It recognizes
that access to oil is no longer simply a matter of seizing it militarily
so as to meet domestic needs—there soon will simply not be enough
to fuel current levels of the industrial output of all nations, no matter
how supplies are distributed. It tacitly recognizes that this inability to
produce enough to satisfy global demand will lead to the escalation of
inter-imperialist rivalry, here blandly described as "competition." It
also recognizes that peak oil constitutes a serious threat to the ability
of the United States to project power.

Globalization, Neoconservatism, and the End of Empire

The antiglobalization movement helped derail the Millennial Round,
forced public meetings behind increasingly militarized police lines,
and compelled the WTO and other international institutions to make
poverty reduction a priority—at least in their rhetoric. Global opposition to neoliberalism, the policy of an unchallenged superpower,
prevented the United States from cementing and normalizing its dominance through the ostensibly peaceful means of trade policy. The
attacks on the World Trade Center and the Pentagon briefly derailed
the antiglobalization movement, and enabled the Bush administration to implement foreign and domestic policies long coveted by the
neoconservative movement. However, although the United States
remains a formidable military power, it is now acting from a position of economic impotence. Rather than marking the beginning of a
"New American Century," neoconservatism was the death rattle of
an empire on life support since the 1970s.

Amid the catastrophic consequences of neoliberal policy upon
the working people (underemployment resulting from capital flight,
growing inequality, and the collapse of social services), and with U.S.
dominance challenged by an international movement calling into
question neoliberal policy and the police and ideological repression
necessary to foist it upon unwilling publics, September 11th offered
the Bush administration the perfect pretext to clamp down on domestic dissent, and to launch a final bid to dominate the distribution of

the oil and natural gas in the Middle East and South Asian region. The PATRIOT Act, clearly prepared well in advance of 9/11, legalized and extended the police tactics employed against the antiglobalization movement. And in the name of bringing the former CIA asset Osama bin Laden to justice, the United States initiated a campaign to subdue Afghanistan. 9/11 provided what the neoconservative Project for the New American Century called "a new Pearl Harbor," unifying a terrified U.S. public to an extent unprecedented in its history.

Although the war against the Taliban regime was generally accepted at face value, opponents questioned the motives of this conflict. Although itself lacking significant fossil fuel reserves, Afghanistan stood in the path of a proposed pipeline to transport natural gas from the Caspian Sea Basin destined, in part, to an Enron electricity generation facility in Dabhol, on the western coast of India. Condemnation of the Afghani government's human rights violations had led the Clinton Administration to prohibit economic cooperation with the Taliban—although its inability to safeguard the proposed pipeline from attack also contributed to waning U.S. support. Plans for an October 2001 invasion were prepared months before 9/11 (Brisard and Dasquie 2002). It soon became clear that the search for bin Laden was halfhearted at best. The installation of Hamed Karzai, a former consultant to one of the member corporations in the consortium that had been pushing for the proposed natural gas pipeline—Unocal—confirmed that U.S. motives had little to do with democracy, sovereignty, or the protection of human rights.

Although the invasion contributed only to the reemergence of warlordism throughout most of the country, the Bush Administration soon launched a massive public relations effort to shift the offensive to its preferred target of Iraq (Rampton and Stauber 2003). This campaign was no more successful than that against Afghanistan. The resistance of the Iraqi people to the occupation is, of course, the proximate cause of the failure of U.S. designs. Life under occupation quickly degenerated, even from the already precarious conditions brought about by the imposition of economic sanctions in the 1990s (Parenti 2004). Best estimates indicate that during the first three years of the war there were about 650,000 deaths resulting from the U.S. invasion and occupation (Burnham et al. 2006), and United Nations High Commission for Refugees estimates that 2 million Iraqis have been internally displaced, with another 2.2 million forced to flee their country altogether (Cockburn 2007). For the Iraqi people, the motives behind this aggression are unmistakable. In reference to a proposed law dictating the future disposition of Iraqi oil—the passage of which was designated a "benchmark of progress" by leaders in both the Republican and Democratic Parties—the General Secretary of the Iraqi Federation of Oil Unions, Faleh Abood Umara, pointed out:

> [C]ontrol of all oil royalties and production [will be given] to foreign oil companies. It will allow them to do whatever they want in our oil fields, and we won't have the ability to intervene, or even to observe.... The law will rob Iraq of its main resource: it's [sic] oil. It will undermine the sovereignty of Iraq and our people. (Bacon 2007)

Although the corporate media in the United States were for a time able to convince most Americans that Iraq was a battleground in the "war on terror," the rush to war was broadly condemned by European allies, and by tens of millions of people through global demonstrations. As the atrocities of Guantanamo and Abu Ghraib, the lies about Iraqi chemical and biological weapons capabilities, and the wholesale carnage let loose upon the Iraqi people made the nature of this war glaringly obvious, the world became increasingly united in opposition to U.S. foreign policy.

A July 2007 survey of five European countries found that 32 percent of respondents believed that the United States government is the greatest threat to world peace. More remarkable is that 11 percent of Americans—and 35 percent of Americans aged 16 to 24—believed that the United States threatens global stability more than any other nation (Dombey and Pignal 2007). Moreover, 78 percent of Americans said they opposed the war against Iraq, 45 percent of Americans reported that they wanted the U.S. House of Representatives to begin impeachment proceedings against the then president George W. Bush, and 54 percent favored the impeachment of the then vice president Dick Cheney (Agence France-Presse 2007). Although mainstream American politicians and media continued to lag far behind the American people, the "Bush Doctrine" of preemptive warfare was thoroughly rejected.

With the focus on Iraq, the United States lost control of what was assumed to be a well-secured portion of its empire—Latin America. Recent elections in Argentina, Bolivia, Brazil, Chile, Ecuador, and Nicaragua were carried by candidates who, to varying degrees, rejected the neoliberal free-trade regime promoted by the U.S. government, were explicitly critical of the neocons, and promised to use the wealth of their nations to eliminate the poverty imposed by the long history of U.S. intervention. Cuba, having recovered from the loss of Soviet aid in the early 1990s, remains a source of inspiration for progressive leaders and movements throughout the world. Venezuelan President Hugo Chavez, having won two elections, having prevailed in two referenda and a recall drive, and having survived two U.S.-supported coup attempts, is nevertheless portrayed by mainstream U.S. politicians and media as on the fringe of world opinion (Rendall, Ward, and Hall 2009).

More ominous is the reemergence of superpower rivalry. Both neo-liberalism and neoconservatism were born of the attempt to project American "leadership" into the unforeseeable future. However, China has absorbed much of the U.S. manufacturing base, and become our pawnbroker. A resurgent Russia successfully ignored hypocritical U.S. protestations of its violation of Georgia's sovereignty, continues to leverage Europe's dependence on its natural gas in its disputes with the former Soviet Republics of Belorus and Ukraine, and appears to have successfully kicked the United States out of a strategic base in Kyrgyzstan (BBC News 2009). Moreover, regional rivalries have become more urgent, as Pakistan's role in both Afghani and Indian terrorism becomes increasingly impossible to ignore, and as Israel's occupation of Palestine turns ever more genocidal.

Imperialism produces contradictory effects, and there can be no "American exceptionalism" with regard to this law of social history. A mere half-century of United States domination was made possible by its oil wealth, but since 1970 it has been increasingly reliant upon the petroleum resources of others. The attempt to dictate the socio-economic arrangements of other peoples—whether through neolib-eral globalization or neoconservative unilateralism—has generated increasing opposition. The neoliberal free market regime decimated the United States' manufacturing capacity, and the attempt to crush resistance continues to require the expenditure of half of the federal government's discretionary budget on its armed forces. Maintaining the world's preeminent military machine amid the erosion of its tax base has entailed massive deficit spending. Thus far, given the status of the U.S. dollar as the world's reserve currency, and especially its status as the medium of exchange for all oil transactions, other nations have been willing to finance these deficits. But this informal arrangement, dubbed "Bretton Woods 2," is unsustainable and has quite probably collapsed (Roubini and Setser 2005, were especially prescient in this regard). Simply put, other nations, most notably China, are loaning the United States government the money that enables it to project military power into the Middle East in an attempt to control these other nations' access to oil. For some time now, the United States has been funding its military machine on installments.

As China and India vie for Middle East and Caspian Basin fossil fuels, and with Russia emerging as a major supplier of oil and natural gas to Europe, the political struggle over these finite resources is again taking center stage. Driven by rapidly growing Asian economies and by stagnant production, the International Energy Agency anticipated that rising demand and tight supplies will lead to soaring oil prices over the next five years (Moore 2007). Although a collapse of the dollar would devalue the holdings of all nations, reserve banks and investors worldwide began to move toward the Euro as support for U.S. fiscal and foreign policies eroded. In October 2000, the Iraqi government demanded compensation for its oil in Euros, and returning to dollar pricing was one of the first effects of the occupation. Iran has made moves toward establishing an oil bourse that would compete with those of the United States and England, and in 2006 China opened its own petroleum exchange. Iran, Russia, Venezuela, and other oil exporters propose trading oil in Euros, Rubles, or through a "basket of currencies" (Clark 2006). As the dollar loses its monopoly status as the medium of exchange for oil, the dollar will become valueless as a reserve currency.

Over the course of early 2008, the price of oil dominated economic news. During the run up that spring, and given the role of fuel in agricultural production, processing, and transport, food prices were held closely in tow. Riots ensued worldwide, and major rice producers chose to abandon free trade and prohibit exports so as to ensure domestic supply and quell political unrest (*The Observer*, 2008). At one point, even retail outlets in the United States placed quotas to prevent runs on rice, beans, and flour (*Seattle Times*, 2008). When the price of oil peaked at $147 a barrel in July, consumer demand in the United States quickly collapsed, with auto purchases leading the way. Although the Ponzi-scheme capitalism of Wall Street helped assemble a house of cards, the bubble in energy and food prices was the proximate tripwire for the economic collapse.

Although the early signs of energy deflation were a welcome respite for many in the United States, the loss of revenue to exporting nations can only contribute to instability in already volatile societies. Regardless of the speed of any economic recovery, supply concerns will quickly reemerge. Oil that was profitable to produce when at $100 a barrel is not profitable to produce at $40 a barrel. As a result, many new projects would be on hold even if credit markets weren't frozen. Any "recovery"—that is, a return to growth in consumption as usual—is bound to generate a new spike in oil prices and a new and deeper destruction of demand.

For progressives worldwide, a bright spot amongst all this has been the election of Barack Obama. As a repudiation of the Bush regime, and indeed as a culmination of the harnessing of antiglobalization grassroots Internet activism, it does indeed bring a glimmer of hope. Without question, international public opinion suggests that many are caught between a sigh of relief at the defeat of McCain, and a collective holding of the breath as Obama's policies unfold. In the United States there appears to be a new willingness to confront the realities of our recent past—support for impeachment has morphed into support for trials of Bush Administration officials, and into a proposal by Senator Patrick Leahy of Vermont for a South African-style Truth Commission. Still, the "fundamentals" haven't changed, and the central policy goal of the new Administration seems to be to restore *the status quo ante*. Judging from the new administration's cabinet, a bipartisan effort to rearrange deck chairs seems in the offing.

Hope, no matter how audacious, won't change the laws of physics, or the laws of capitalist development. Class analysis has long forecast this crisis of capitalist imperialism. A confluence of dire predictions, from a range of perspectives, should give pause to any naïve optimism. Although covered by U.S. media only rarely, and in derision, analysts are predicting not only the loss of U.S. influence internationally and a deep and prolonged depression, but the "de-development" of industrial societies, and even the breakup of the continental United States itself (Gatinois 2009; Kunstler 2005; Osborn 2008). Whatever the details of the future, it is increasingly clear that U.S. domination of the world ended quite some time ago. Moreover, it seems reasonable to suggest that since American society has been predicated on empire, the end of empire will entail a fundamental restructuring of American society.

Conclusion

With the groundwork laid by European colonialism and imperialism, a pattern of international relationships dominated by the United States emerged in the early twentieth century, and it was institutionalized after World War II. With the dissolution of the Soviet Union, the United States attempted to extend its control of the world and push the empire's frontiers as far as possible. In the face of effective global opposition to "democratic" neoliberalism, and in light of the urgency of peak oil, the United States government turned to a policy of unapologetic intervention and occupation. The crimes of empire, the dismissal of the concerns of U.S. allies and official enemies alike, and the futility of a New American Century amid the end of the Age of Oil, only served to isolate the United States, to embolden rival empires, and to energize popular opposition.

In the immediate aftermath of the attacks on the World Trade Center and Pentagon, many expected that the tragedy would serve to deflate progressive activism. But organizations and coalitions that were nurtured and tempered in the antiglobalization movement made important contributions to a new turn in popular awareness of, and opposition to, neoliberalism, neoconservatism, and imperialism.

U.S. foreign policy has unleashed chaos and misery around the world, and there is no question that the detritus of these futile attempts at global domination will continue to maim and kill future generations no matter how aggressively and unselfishly we might recommit ourselves today. Even if, overnight, all people were freed from foreign domination, all nations were to adopt egalitarian and otherwise humane policies and practices, every effort were made to eradicate poverty, promote the benefits of democratic governance, and we were all to spare no expense in cleaning up the flotsam and jetsam of modern civilization and in developing sustainable economies and communities, untold generations will still die from hunger and squalor, will lose their homelands to rising oceans and desertification, will plow fields with unexploded ordinance, and will suffer deformities and cancers as a result of the literally toxic sociopathy of our time.

But, as a direct result, most people have come to recognize that creating a world fit for humans will require that we reject empires and imperialism, however euphemized under the cloak of neoliberal globalization, and embrace cooperation and consensus in place of violence and self-destructive competition for private accumulation. This transition has been and will continue to be painful, and will require continued struggle against all forms of exploitation and oppression. Nevertheless, the path ahead seems clear, and the means by which the anti-imperialist forces will ultimately prevail appear to be at hand.

References

Ackerman, Seth. 2000. "Prattle in Seattle: WTO Coverage Misrepresented Issues, Protests." *Extra!* January/February: 13.

Agence France-Presse. 2007. "Much of US Favors Bush Impeachment: Poll," accessed online at *http://www.afp.com/english/news/stories/070706225930. sv6rw08w.html.*

Allen, Terry. 2000. "Chemical Cops." *In These Times,* 24, April 3: 14–17.

Bacon, David. 2007. "Iraqi's Oil Should Stay in Public Hands." *Truthout.org,* accessed online at *http://www.truthout.org/docs_2006/071007C.shtml*

Bagdikian, Ben. 2004. *The Media Monopoly.* Boston: Beacon Press.

Baker, Dean, 2000. "Free Trade Fables: Economic Misinformation in WTO Coverage." *Extra!* January/February: 18.

BBC News. 2009. "Kyrgyz MPs Vote to shut U.S. Base" (February 2009), accessed online at *http://news.bbc.co.uk/2/hi/asia-pacific/7898690.stm*

Brisard, Jean Charles and Guillame Dasquie. 2002. *Forbidden Truth: U.S.- Taliban Secret Oil Diplomacy and the Failed Hunt for bin Laden,* translated by Lucy Rounds. New York: Thunder's Mouth Press.

British Petroleum. 2007. "Statistical Review of World Energy," accessed online at *http://www.bp.com/liveassets/bp_internet/globalbp/globalbp_uk_eng-lish/reports_and_publications/statistical_energy_review_2007/STAGING/ local_assets/downloads/pdf/statistical_review_of_world_energy_full_ report_2007.pdf.*

———. 2008. "Statistical Review of World Energy," accessed online at *http:// www.bp.com/liveassets/bp_internet/globalbp/globalbp_uk_english/ reports_and_publications/statistical_energy_review_2008/STAGING/ local_assets/downloads/pdf/statistical_review_of_world_energy_full_ review_2008.pdf.*

Burnham, Gilbert, Riyadh Lafta, Shannon Doocy, and Les Roberts. 2006. "Mortality after the 2003 Invasion of Iraq: A Cross-sectional Cluster Sample Survey." *The Lancet,* accessed online at *http://www.thelancet.com/webfiles/ images/ journals/lancet/s0140673606694919.pdf/*

Campbell, Colin. 2004. *The Coming Oil Crisis.* Brentwood, UK: Multi-Science Publishing Company, Ltd.

Chomsky, Noam. 2003. *Hegemony or Survival: The Imperialist Strategy of the United States.* New York: Metropolitan Books.

Clark, William. 2006. "Hysteria over Iran and a New Cold War with Russia: Peak Oil, Petrocurrencies and the Emerging Multipolar World," accessed online at *http://globalresearch.ca/articles/williamclarkrussia.pdf.*

Cockburn, Patrick. 2007. "UN Warns of Five Million Iraqi Refugees." *The Independent,* July 17, accessed online at *http://news.independent.co.uk/ world/middle_east /article2640418.ece*

Coen, Rachel. 2000. "For Press, Magenta Hair & Nose Rings Defined Protests." *Extra!* July/August: 10.

———. 2000a. "Whitewash in Washington: Media Provide Cover as Police Militarize D.C." *Extra!* January/February: 9–12.

Compaine, Benjamin M. and Douglas Gomery. 2000. *Who Owns the Media? Competition and Concentration in the Mass Media Industry,* 3rd edition, Mahwah, NJ: Lawrence Erlbaum Associates.

Darley, Julian. 2004. *High Noon for Natural Gas: The New Energy Crisis.* White River Junction: Chelsea Green Publishing.

Deffeyes, Kenneth. 2001. *Hubbert's Peak: The Impending World Oil Shortage.* Princeton, NJ: Princeton University Press.

Dombey, Daniel and Stanley Pignal. 2007. "Europeans See US as Threat to Peace." *The Financial Times,* July 1, accessed online at *http://www.ft.com/cms/s/70046760-27f0-11dc-80da-000b5df10621.html.*

Engdahl, F. William. 1993. *A Century of War: Anglo-American Oil Politics and the New World Order,* 1st English edition. Concord, MA: Paul and Company Publishing Consortium.

Flanders, Laura. 2000. "Move Over, Pacifica! Here Come Indy Media." *CounterPunch,* April 15–30: 7.

Gatinois, Claire. 2009. "After the Financial Crisis, Civil War? Get Ready to 'Leave Your Region.'" *Le Monde,* translation by Leslie Thatcher, February 26, accessed online at http://www.truthout.org/022709F.

Haski, Pierre. 2000. "The Bosses Grow Rich on Vocabulary." *Libération,* January 31, reprinted in *World Press Review,* April: 9.

Heinberg, Richard. 2003. *The Party's Over: Oil, War, and the Fate of Industrial Societies,* Gabriola Island, British Columbia: New Societies Publishers.

———. 2004. *Powerdown: Options and Actions for a Post-carbon World.* Gabriola Island, British Columbia: New Societies Publishers.

Herman, Edward S. and Noam Chomsky. 1988. *Manufacturing Consent: The Political Economy of the Mass Media.* New York: St. Martins.

Higham, Charles. 1983. *Trading with the Enemy: An Exposé of the Nazi-American Money Plot 1933–1949.* New York: Delacorte Press.

Klare, Michael T. 2001. *Resource Wars: The New Landscape of Global Conflict.* New York: Henry Holt and Company.

———. 2009. *Rising Powers, Shrinking Planet: The New Geopolitics of Energy.* New York: Henry Holt and Company.

Kunstler, James Howard. 2005. *The Long Emergency: Surviving the Converging Catastrophes of the Twenty-First Century.* New York: Atlantic Monthly Press.

Leibovitz, Clement and Alvin Finkel. 1997. *In Our Time: The Chamberlain-Hitler Collusion.* New York: Monthly Review Press.

McChesney, Robert W. 2008. *The Political Economy of Media: Enduring Issues, Emerging Dilemmas.* New York: Monthly Review Press.

McDonald, Colin. 2007. "WTO Protesters to Receive $1 Million: City's Settlement is Largest and Last to Arise from Suits." *Seattle Post-Intelligencer,* April 2, accessed online at *http://seattlepi.nwsource.com/local/310008_wto03.html.*

Moore, James. 2007. "Energy Watchdog Warns of Supply Crunch within Five Years." *The Independent UK,* July 10, accessed online at *http://news.independent.co.uk/ business/news/article2750517.ece.*

Morales, Frank. 2000. "'Emergency Management' in Seattle." *Covert Action Quarterly,* Spring/Summer: 63.

The Observer. 2008. "Food Riots Fear After Rice Price Hits a High," April 9, accessed online at *http://www.guardian.co.uk/environment/2008/apr/06/food.foodanddrink.*

Osborn, Andrew. 2008. "As If Things Weren't Bad Enough, Russian Professor Predicts End of U.S.: In Moscow, Igor Panarin's Forcasts are All the Rage; America 'Disintegrates' in 2010." *The Wall Street Journal,* December 29: A1

Parenti, Christian. 2004. *The Freedom: Shadows and Hallucinations in Occupied Iraq*. New York: The New Press.

Parenti, Michael. 1986. *Inventing Reality: The Politics of the Mass Media*. New York: St. Martins.

———. 1997. *Blackshirts and Reds: Rational Fascism and the Overthrow of Communism*. San Francisco: City Lights Books.

Rampton, Sheldon and John Stauber. 2003. *Weapons of Mass Deception: The Uses of Propaganda in Bush's War on Iraq*. New York: Tarcher/Penguin.

Rendall, Steve, Daniel Ward and Tess Hall. 2009. "Human Rights Coverage Serving Washington's Needs." *Extra!* February:7–10.

Roubini, Nouriel and Brad Setser. 2005. "Will the Bretton Woods 2 Regime Unravel Soon? The Risk of a Hard Landing in 2005–2006." Paper written for the Symposium on the Revived Bretton Woods System: A New Paradigm for Asian Development? organized by the Federal Reserve Bank of San Francisco and the University of California, Berkeley, accessed online at *http://pages. stern.nyu.edu/ ~nroubini/papers/BW2-Unraveling-Roubini-Setser.pdf*.

Scher, Abby. 2001. "The Crackdown on Dissent." *The Nation*, February 5: 23–24.

The Seattle Times. 2008. "Rice Hoarding: The Long and Short of It," April 29, accessed online at*http://seattletimes.nwsource.com/html/editorialsopinion/ 2004379459_riced29.html*

Seldes, George. 1943. *Facts and Fascism*. New York: In Fact, Inc.

Simmons, Matthew R. 2005. *Twilight in the Desert: The Coming Saudi Oil Shock and the World Economy*. Hobokin, NJ: John Wiley & Sons, Inc.

Simpson, Christopher. 1993. *The Splendid Blonde Beast*. New York: Grove.

Sunde, Scott. 1999. "Chaos Closes Downtown." *The Seattle Post-Intelligencer*, December 1: A1.

Trofimov, Yaroslav and Ian Johnson. 2001. "G-8 Protesters in Italy Describe Police Attack on Group in a School: In Interviews, They Say They Were Beaten, Deprived of Rights in Detention." *The Wall Street Journal*, August 6: A1, A6.

United States Army Corp of Engineers. 2005. "Energy Trends and Implications for U.S. Army Installations," accessed online at *www.peakoil.net/Articles2005/ Westervelt_ EnergyTrends__TN.pdf*.

Weissman, Robert. 1999. "Welcome to Seattle: Ministerial Meeting Debates the World Trade Organization's Agenda for the 21st Century." *Multinational Monitor*, October/November: 9–13.

World Trade Organization. 2007. "Understanding the WTO," accessed online at *http://www.wto.org/English/thewto_e/whatis_e/tif_e/org6_e.htm*.

Yeoman, Matthew. 2004. *Oil: A Concise Guide to the Most Important Product on Earth*. New York: The New Press.

Zachary, G. Pascal. 2000. "The New Imperialism: A Year After the 'Battle in Seattle,' the Rich are Embracing Their Critics. Watch out." *In These Times*, December 25: 30.

10

Conclusion: Globalization for a New Century

Berch Berberoglu

It has been argued throughout this book that neoliberal globaliza-tion—much celebrated in recent decades for ushering in a period of "prosperity for all"—has come under serious criticism and opposition from movements around the world that have waged (and continue to wage) a successful struggle against it. Movements made up of work-ers, peasants, indigenous peoples, environmentalists, intellectuals, students, and others have come together in a broad coalition of forces to oppose corporate globalization that has advanced the interests of the dominant classes in society to profit from the exploitation and oppression of the great majority of the people, while forcing millions into poverty and destitution. From sweatshop slavery to global warm-ing, from mass unemployment to military interventions, and from speculative profiteering to ethnic cleansing, the capitalist powers have used the neoliberal cover to justify every shady imperialist activity to the benefit of their global empire that is now crumbling in the face of the worst economic collapse since the Great Depression. Thus, the era of neoliberalism, that is, of neoliberal capitalist globalization, is coming to an end. And people across the globe are coming together to form a powerful coalition of forces to resist the global capitalist system and reverse its course in a last-ditch effort to save the planet from those who are intent on destroying it.

Whereas the twentieth century was the era of capitalist imperialism that developed into neoliberal globalization (i.e., global capitalism in sheep's skin), the devastation and havoc that it has caused the world over is now being accounted for—and the record is very grim. Movements around the world, led by working people and their class-conscious

organizations in alliance with a variety of popular forces, are pro-
claiming that "a new world is possible" and *necessary* to reverse the
tide of total catastrophe that is waiting to happen. And it is with this
hope, based on the hard work of dedicated people wanting to build an
equitable and a just world, that one can see the emergence of a post-
neoliberal-capitalist world order that will have a diversity of forces to
promote popular democratic principles that serve the greater good.

Decline of empire and superpower domination of the global political economy through its own dynamics and contradictions is also
setting into motion a process by which new multiple power centers
independent of the center states are bound to emerge. This signals
the rise of new forces on the global scene in the coming years that
have been in the background during the past several decades. Their
entry into the global system and prominence in it in future years will
confirm the political and economic shifts that the world is now beginning to experience in a new way. It is within the context of these
newly emergent forces that we will see the development of the global
economy and polity in the direction of multipolarity in the new twenty-first century. This process will inevitably be more decentralized,
more democratic, and more equitable across the globe. But it is not
a process that will unfold on its own accord through some imper-
sonal forces of nature. In fact, it would be the end result of human
social action through collective political will that demands social justice. It is through this kind of effort in organized political action that
people's movements and their allies would be able to effect massive
systemic change.

The evolution, development, and transformation of capitalism from
the twentieth to the twenty-first century is of epochal significance, as
the heretofore chief actors of the capitalist system go on the defensive to restructure the moribund and decaying neoliberal capitalist
economy through state intervention and regulation to save the system
from itself, as it self-destructs through a series of deep and long-last-
ing recessions and depressions stemming from global financial crises.
Thus, as the epoch of neoliberal transnational capitalism comes to an
end, many are wondering what will come to replace it. As in the case
of the bankrupt capitalist corporations, will the new global economy
emerge as a leaner and more "efficient" reincarnation of its old self
in a new form on the world scene? Or will it die the way of ENRON,
WorldCom, Lehman Brothers, AIG, Citi Group, Chrysler, General
Motors, and a dozen other icons of twentieth-century capitalism that
have vanished or are about to vanish into the dustbin of history? Or,

will global capital be given a new lease of life by upstart noncapitalist forces that will come to the rescue of capital while building their own power base as future superpowers of unknown recent (and future) credentials that call into question the entire global enterprise from which they may emerge as a powerful force and are now destined to replace it? We are, of course, referring in particular to China as the next giant to play a decisive role in the shaping and reshaping of the global economy in the twenty-first century.

The global economic crisis has affected the capitalist world as well as those that have integrated into it from outside the system in a big way, as they rely on their trade and investment priorities that are associated with the global capitalist system. Whereas this is true of China, Russia, and other traditionally socialist states that have recently joined the capitalist global economy to advance their own interests (and the nature of these interests are not precisely known because of their complex character through heavy state intervention by bureaucratic forces within a capitalist global environment), the current global economic crisis, dampening the once-vibrant transnational trade, may force some of these states, like China and Russia, to turn inward and reenergize their internal market. As consumers in these societies increase their purchasing power through a rise in the standard of living propelled by internal mechanisms of industrial growth and increasing wages and benefits instituted by the state, the growing power of well-off workers as consumers could trigger a new cycle of national economic boom in these countries while the rest of the world struggles to get out of the economic crisis in which they have been trapped. This could almost certainly mean the emergence of China and Russia (separately or combined) as the next superpower(s) in the global economy over the course of the next decade, if the continuing decline and collapse of the global capitalist economy does not lead to another major world war, targeting these two emergent powers that may challenge the very basis of the global capitalist system. Thus, while withdrawing from the global economy may, by necessity, signal the rise of China as the next giant economic (and ultimately military) superpower by the middle of the twenty-first century, the epochal contradictions and crises of global capitalism places the future of the planet at great risk through another colossal world war.

Globalization in the twenty-first century is surely going to look different than it did in the twentieth century, and it is the nature and scope of developments that we are now undergoing in the early twenty-first century that will set the stage for changes that are yet to come in the decades ahead. As has been suggested in a number of essays comprising this book, we are moving toward a multipolar, multilayered complex arrangement of competing centers in the global political economy that will counterbalance the interests of the old powers with those of the new, while conflicts and contradictions of these competing states will get resolved through their external relations and internal class dynamics as they come to experience many of the inherent tensions that are built into these societies because of their very nature and structure.

Twenty-first-century globalization will definitely be more decentralized, more democratic, and more socially oriented, as well as prosperous, dynamic, and forward-looking. These changing qualities of globalization in the new century will be the product of the mass movements struggling against neoliberal capitalist globalization, which will be transformed and replaced with a higher, superior global world order beyond capitalism, imperialism, and neoliberal globalization that has meant much misery for the masses throughout the world. And this raises questions regarding the level of class consciousness of working people and their allies, their organizational skills, their ideological sophistication, and their willingness to take political action through a revolutionary process that comes to mobilize the masses for decisive battles to win the wars that are necessary to take power and rule society. If the previous rulers of the world were the capitalists, the capitalist states, and their global capitalist institutions that facilitated and assured the rule of the capitalist class over the globe, now controlling the commanding heights of the economy, polity, and society, it is labor's turn to take charge and construct a new global social order that inaugurates for the first time in history a worldwide effort to bring justice and equity to the lives of billions of working people around the globe through the principles and practices of a new social, economic, and political system.

It is through this hope for a better world, which is possible through human will based on the value placed on labor, that we stand a chance to reverse the disastrous course in which people have been the victims throughout human history, and play an active role to set it straight through concerted human action with the efforts of millions of dedicated people who are destined to save the future of humanity. It is, to be sure, the working class and its class-conscious organizations that will come to the rescue by pushing aside the collapsing, crisis-ridden capitalist world order and replacing it with socialism to transform society in the direction of a truly human existence in the new twenty-first century.

Bibliography

Amoroso, Bruno. 2001. *On Globalization: Capitalism in the 21st Century.* New York: St. Martins.

Amaladoss, Michael (ed.). 1999. *Globalization and Its Victims As Seen by Its Victims.* Delhi, India: Vidyajyoti Education and Welfare Society.

Andor, Laszlo and Martin Summers. 1998. *Market Failure: A Guide to the East European "Economic Miracle."* London: Pluto.

Appadurai, Arjun. 2001. *Globalization.* Durham, NC: Duke University Press.

Arrighi, Giovanni. 1994. *The Long Twentieth Century.* London: Verso.

Arrighi, Giovanni and Beverly Silver. 1999. *Chaos and Governance in the Modern World-System.* Minneapolis: University of Minnesota Press.

Ayala, Cesar J. 1989. "Theories of Big Business in American Society." *Critical Sociology* 16, 2–3 (Summer–Fall).

Bacon, David. 2004. *The Children of NAFTA: Labor Wars on the U.S./Mexico Border.* Berkeley, CA: University of California Press.

Bamyeh, Mohammed A. 2000. *The Ends of Globalization.* Minneapolis: University of Minnesota Press.

Barkin, David, Irene Ortiz, and Fred Rosen. 1997. "Globalization and Resistance: The Remaking of Mexico." *NACLA* 30, 40.

Barndt, Deborah (eds.). 1999. *Women Working the NAFTA Food Chain: Women, Food & Globalization.* Toronto: Sumach.

———. 2002. *Tangled Routes: Women, Work, and Globalization on the Tomato Trail.* Lanham, MD: Rowman and Littlefield.

Barnet, Richard and John Cavenagh. 1994. *Global Dreams: Imperial Corporations and the New World Order.* New York: Simon & Schuster.

Bauman, Z. 1998. *Globalization: The Human Consequences.* Cambridge, UK: Polity.

Baylis, John, Steve Smith, and Patricia Owens. 2008. *The Globalization of World Politics.* 4th ed. New York: Oxford University Press.

Beams, Nick. 1998. *The Significance and Implications of Globalization: A Marxist Assessment.* Southfield, UK: Mehring.

Bello, Walden. 1994. *Dark Victory: United States, Structural Adjustment and Global Poverty.* Herndon, VA: Pluto.

Bello, Walden and Stephanie Rosenfeld. 1990. *Dragons in Distress: Asia's Miracle Economies in Crisis.* San Francisco: Institute for Food and Development Policy.

Benería, Lourdes. 1989. "Gender and the Global Economy." In *Instability and Change in the World Economy*, ed. Arthur MacEwan and William K. Tabb. New York: Monthly Review Press.

———. 2003. *Gender, Development, and Globalization*. New York: Routledge.

Berberoglu, Berch. 1987. *The Internationalization of Capital: Imperialism and Capitalist Development on a World Scale*. New York: Praeger.

———. 1992a. *The Legacy of Empire: Economic Decline and Class Polarization in the United States*. New York: Praeger.

———. 1992b. *The Political Economy of Development: Development Theory and the Prospects for Change in the Third World*. Albany: SUNY Press.

———. 2001. *Political Sociology: A Comparative/Historical Approach*. 2nd ed. New York: General Hall.

———. 2002. *Labor and Capital in the Age of Globalization: The Labor Process and the Changing Nature of Work in the Global Economy*. Lanham, MD: Rowman and Littlefield.

———. 2003. *Globalization of Capital and the Nation-State: Imperialism, Class Struggle, and the State in the Age of Global Capitalism*. Lanham, MD: Rowman and Littlefield.

———. 2005. *Globalization and Change: The Transformation of Global Capitalism*. Lanham, MD: Lexington Books.

———. 2007. *The State and Revolution in the Twentieth Century: Major Social Transformations of Our Time*. Lanham, MD: Rowman and Littlefield.

———. 2009. *Class and Class Conflict in the Age of Globalization*. Lanham, MD: Lexington Books.

Berger, Peter and Samuel P. Huntington (eds.). 2002. *Many Globalizations: Cultural Diversity in the Contemporary World*. New York: Oxford University Press.

Bergman, Gregory. 1986. "The 1920s and the 1980s: A Comparison." *Monthly Review* 38, 5 (October).

Berik, Günseli. 2000. "Mature Export-Led Growth and Gender Wage Inequality." *Feminist Economics* 6 (3): 1–26.

Beynon, John and O. Dunkerley (eds.). 2000. *Globalization: The Reader*. New York: Routledge.

Bina, Cyrus and Behzad Yaghmaian. 1991. "Post-war Global Accumulation and the Transnationalization of Capital." *Capital & Class*, 43 (Spring).

Bina, Cyrus and Chuck Davis. 1996. "Wage Labor and Global Capital: Global Competition and the Universalization of the Labor Movement." In *Beyond Survival: Wage Labor and Capital in the Late Twentieth Century*, ed. Cyrus Bina, Laurie Clements, and Chuck Davis . Armonk, NY: Sharpe.

———. 2000. "Globalization, Technology, and Skill Formation in Capitalism." In *Political Economy and Contemporary Capitalism: Radical Perspectives on Economic Theory and Policy*, ed. Ron Baiman, Heather Boushey, and Dawn Saunders. Armonk, NY: Sharpe.

———. 2002. "Dynamics of Globalization: Transnational Capital and the International Labor Movement." In *Labor and Capital in the Age of Globalization: The Labor Process and the Changing Nature of Work in the Global Economy*, ed. Berch Berberoglu. Lanham, MD: Rowman and Littlefield.

Black, Stanley W. (ed.). 1998. *Globalization, Technological Change, and Labor Markets*. Boston: Kluwer.

Bluestone, Barry and Bennett Harrison. 1982. *The Deindustrialization of America*. New York: Basic.

Blum, William. 2001. *Rogue State: A Guide to the World=s Only Superpower*. London: Zed.

Bonacich, Edna, Lucie Cheng, Norma Chinchilla, Nora Hamilton, and Paul Ong (eds.). 1994. *Global Production: The Apparel Industry in the Pacific Rim*. Philadelphia: Temple University Press.

Borrego, John, Alejandro Alvarez Bejar, and Jomo K. S. (eds.). 1996. *Capital, the State, and Late Industrialization*. Boulder, CO: Westview.

Bowles, Samuel and Richard Edwards. 1985. *Understanding Capitalism*. New York: Harper & Row.

Boyd-Barrett, Oliver and Terhi Rantanen (eds.). 1998. *The Globalization of News*. Thousand Oaks, CA: Sage.

Braun, Denny. 1991. *The Rich Get Richer: The Rise of Income Inequality in the United States and the World*. Chicago: Nelson-Hall.

Brecher, Jeremy and Tim Costello. 1998. *Global Village or Global Pillage: Economic Reconstruction from the Bottom Up*. Cambridge, MA: South End.

Brecher, Jeremy, Tim Costello, and Brendan Smith. 2000. *Globalization From Below: The Power of Solidarity*. Cambridge, MA: South End.

Brenner, Robert. 1977. "The Origins of Capitalist Development: A Critique of Neo-Smithian Marxism." *New Left Review* 104.

Bresheeth, Haim and Nira Yuval-Davis (eds.) 1991. *The Gulf War and the New World Order*. London: Zed.

Broad, Robin. 2002. *Global Backlash: Citizen Initiatives for a Just World Economy*. Lanham, MD: Rowman and Littlefield.

Brown, Jonathan C. (ed.). 1997. *Workers' Control in Latin America, 1930–1979*. Chapel Hill: University of North Carolina Press.

Brown, Michael Barratt. 1974. *The Economics of Imperialism*. Baltimore, MD: Penguin.

Bryan, Dick. 1995a. "The Internationalization of Capital and Marxian Value Theory." *Cambridge Journal of Economics*, 19.

———.1995b. *The Chase Across the Globe: International Accumulation and the Contradictions for Nation States*. Boulder, CO: Westview.

Burbach, Roger, Orlando Nunez, and Boris Kagarlitsky. 1996. *Globalization and Its Discontents: The Rise of Postmodern Socialisms*. London: Pluto.

———. 2001. *Globalization and Postmodern Politics*. London: Pluto.

Buxton, Julia and Nicola Phillips (eds.). 1999. *Case Studies in Latin American Political Economy*. Manchester, UK: Manchester University Press.

Calleo, David. 1987. *Beyond American Hegemony: The Future of the Western Alliance*. New York: Basic.

Callinicos, Alex. 1991. *The Revenge of History: Marxism and the East European Revolutions*. University Park: Pennsylvania State University Press.

Cammack, Paul. 1997. *Capitalism and Democracy in the Third World*. London: Leicester University.

Cantor, Daniel and Juliet Schor. 1987. *Tunnel Vision: Labor, the World Economy, and Central America*. Boston: South End.

Cardoso, Fernando Henrique. 2001. *Charting a New Course: The Politics of Globalization and Social Transformation*. Lanham, MD: Rowman and Littlefield.

Carnoy, Martin. 1984. *The State and Political Theory*. Princeton, NJ: Princeton University Press.

Chase-Dunn, Christopher. 1998. *Global Formation: Structures of the World Economy*. Lanham, MD: Rowman and Littlefield.

Chase-Dunn, Cristopher, Susanne Jonas, and Nelson Amaro (eds.). 2001. *Globalization on the Ground*. Lanham, MD: Rowman and Littlefield.

Cheng, Y. S. (ed.). 2007. *Challenges and Policy Programmes of China's New Leadership*. Hong Kong: City University of Hong Kong Press.

Cherry, Robert, Christine D'Onofrio, Cigdem Kurdas, Thomas R. Michl, Fred Moseley, and Michele I. Naples (eds.). 1987. *The Imperilled Economy*, Book I. New York: Union for Radical Political Economics.

Chossudovsky, Michel. 1997. *The Globalization of Poverty*. London: Zed.

Clapp, Jennifer. 2001. *Toxic Exports: The Transfer of Hazardous Wastes from Rich to Poor Countries*. Ithaca, NY: Cornell University Press.

Clark, Gordon L. and Won Bae Kim (eds.). 1995. *Asian NIEs and the Global Economy: Industrial Restructuring and Corporate Strategy in the 1990s*. Baltimore, MD: Johns Hopkins University Press.

Clark, Robert P. 2001. *Global Life Systems: Population, Food, and Disease in the Process of Globalization*. Lanham, MD: Rowman and Littlefield.

———. 2002. *Global Awareness: Thinking Systematically About the World*. Lanham, MD: Rowman and Littlefield.

Clawson, Dan. 2003. *The Next Upsurge: Labor and the New Social Movements*. Ithaca, NY: Cornell University Press.

Cockcroft, James D. 1996. *Latin America: History, Politics, and U.S. Policy*. 2nd ed. Belmont, CA: Wadsworth.

Cohen, Daniel. 2007. *Globalization and Its Enemies*. Cambridge, MA: The MIT Press.

Cohn, Theodore H., Stephen McBride, and John Wiseman (eds.). 2000. *Power in the Global Era: Grounding Globalization*. New York: St. Martin's.

Collins, Susan M. (ed.). 1998. *Imports, Exports, and the American Worker*. Washington, DC: Brookings.

Comeliau, Christian. 2002. *The Impasse of Modernity: Debating the Future of the Global Market Economy*. London: Zed.

Cornwell, Grant H. and Eve Walsh Stoddard (eds.). 2000. *Global Multiculturalism: Comparative Perspectives on Ethnicity, Race, and Nation*. Lanham. MD: Rowman and Littlefield.

Cox, Robert W. 1997. "A Perspective on Globalization." In *Globalization: Critical Reflections*, ed. J. H. Mittelman. Boulder, CO: Lynne Rienner Publishers, 1997.

Cravey, Altha J. 1998. *Women and Work in Mexico's Maquiladoras*. Lanham, MD: Rowman and Littlefield.

Croucher, Sheila L. 2004. *Globalization and Belonging: The Politics of Identity in a Changing World*. Lanham. MD: Rowman and Littlefield.

Cuyvers, Ludo. 2001. *Globalization and Social Development: European and Southeast Asian Evidence*. Cheltenham, UK: Edward Elgar.

Dadush, Uri B., Dipak Dasgupta, and Marc Uzan (eds.). 2001. *Private Capital Flows in the Age of Globalization: The Aftermath of the Asian Crisis*. Cheltenham, UK: Edward Elgar.

Danaher, Kevin, and Roger Burbach (eds.). 2000. *Globalize This! The Battle Against the World Trade Organization and Corporate Rule*. Monroe, ME: Common Courage.

Davis, Mike. 1984. "The Political Economy of Late Imperial America." *New Left Review* 143 (January-February).

De Caux, Len. 1970. *Labor Radical*. Boston: Beacon.

_____. 1974. "UE: Democratic Unionism at Work." *World Magazine* (April).

De la Barra, Ximena and Richard A. Dello Buono. 2008. *Latin America After the Neoliberal Debacle*. Lanham, MD: Rowman and Littlefield.

Della Porta, Donatella and Sidney Tarrow (eds.). 2005. *Transnational Protest and Global Activism*. Lanham, MD: Rowman and Littlefield.

Devine, Jim. 1982. "The Structural Crisis of U.S. Capitalism." *Southwest Economy and Society* 6, 1 (Fall).

Deyo, Frederick, Stephen Heggard, and Hagen Koo. 1987. "Labor in the Political Economy of East Asian Industrialization." *Bulletin of Concerned Asian Scholars* 19, 2 (April-June).

Dicken, Peter. 1992. *Global Shift: The Internationalization of Economic Activity*. New York: Guilford.

Dickenson, Torry D. and Robert K. Schaeffer. 2001. *Fast Forward: Work, Gender, and Protest in a Changing World*. Lanham, MD: Rowman and Littlefield.

Dollar, David. 2001. "Globalization, Inequality, and Poverty Since 1980." Development Research Group. Washington, DC: World Bank.

Dugger, William and Howard J. Sherman. 2000. *Reclaiming Evolution: A Dialogue between a Marxist and an Institutionalist*. London: Routledge.

Dworkin, Dennis. 2007. *Class Struggles*. London: Longman.

Eckstein, Susan (ed.). 2001. *Power and Popular Protest: Latin American Social Movements*. Berkeley: University of California Press.

Edoho, Felix Moses (ed.). 1997. *Globalization and the New World Order: Promises, Problems, and Prospects for Africa in the Twenty-First Century*. Westport, CT: Praeger.

Edwards, Michael and John Gaventa (eds.). 2001. *Global Citizen Action*. Boulder, CO: Rienner.

Eitzen, D. Stanley and Maxine Baca Zinn. 2008. *Globalization: The Transformation of Social Worlds*. 2nd ed. Belmont, CA: Wadsworth.

Elbakidze, Marina (ed.). 2002. *Globalization: A Bibliography with Indexes*. New York: Nova Science.

Emigh, Rebecca Jean and Ivan Szelenyi (eds.). 2000. *Poverty, Ethnicity, and Gender in Eastern Europe During the Market Transition*. New York: Praeger.

Esping-Andersen, Gosta, Roger Friedland, and Erik Olin Wright. 1976. "Modes of Class Struggle and the Capitalist State." *Kapitalistate*, 4–5 (Summer).

Falk, Richard. 1999. *Predatory Globalization: A Critique*. Malden, MA: Blackwell.

Fantasia, Rick. 1988. *Cultures of Solidarity: Consciousness, Action, and Contemporary American Workers*. Berkeley: University of California Press.

Faux, Geoffrey P. 2006. *The Global Class War*. Hoboken, NJ: Wiley.

Fleming, D. F. 1961. *The Cold War and Its Origins, 1917–1960*. 2 vols. New York: Doubleday.

Foran, John (ed.). 2002. *The Future of Revolutions: Rethinking Political and Social Change in the Age of Globalization*. London: Zed.

———. 2005. *Taking Power: On the Origins of Third World Revolutions*. Cambridge: Cambridge University Press.

Foster, John B. 1986. *The Theory of Monopoly Capitalism*. New York: Monthly Review.

———. 2001. "Imperialism and 'Empire.'" *Monthly Review* 53, 7 (December).

———. 2002. "Monopoly Capital and the New Globalization." *Monthly Review* 53, 8 (January).

Foster, John B. and Henryk Szlajfer (eds.). 1984. *The Faltering Economy: The Problem of Accumulation under Monopoly Capitalism*. New York: Monthly Review Press.

French, Hilary F. 2000. *Vanishing Borders: Protecting the Planet in the Age of Globalization*. New York: Norton.

Friedman, Jonathan (ed.). 2002. *Globalization, the State, and Violence*. Lanham, MD: Rowman and Littlefield.

Friedman, Kajsa Ekholm and Jonathan Friedman. 2002. *Global Anthropology*. Lanham, MD: Rowman and Littlefield.

Friedman, Thomas L. 2007. "It is a Flat World After All." In *The Globalization and Development Reader: Perspectives on Development and GlobalChange*, ed. Roberts, J. T. and A. B. Hite. Malden, MA: Blackwell.

Fuentes, Annette and Barbara Ehrenreich. 1983. *Women in the Global Factory*. Boston: South End.

Gerstein, Ira. 1977. "Theories of the World Economy and Imperialism." *Insurgent Sociologist* 7, 2 (Spring).

Giddens, Anthony. 2000. *Runaway World: How Globalization Is Reshaping Our Lives*. New York: Routledge.

Gills, Barry K. 2001. *Globalization and the Politics of Resistance*. New York: Macmillan.

Gold, David, Clarence Y. H. Lo, and Erik Olin Wright. 1985. "Recent Developments in Marxist Theories of the Capitalist State." *Monthly Review* 27, 5–6 (October, November).

Golding, Peter and Phil Harris (eds.). 1997. *Beyond Cultural Imperialism: Globalization, Communication and the New International Order*. Thousand Oaks, CA: Sage.

Gosh, B. N. and Halil M. Guven (eds.). 2006. *Globalization and the Third World: A Study of Negative Consequences*. New York: Palgrave Macmillan.

Green, Felix. 1971. *The Enemy: What Every American Should Know about Imperialism*. New York: Vintage.

Griffen, Sarah. 1991. "The War Bill: Adding Up the Domestic Costs of War." *Dollars & Sense*, 165 (April).

Griffin, Keith. 1995. "Global Prospects for Development and Human Security." *Canadian Journal of Development Studies* 16, 3.

Griffin, Keith and Rahman Khan. 1992. *Globalization and the Developing World: An Essay on the International Dimensions of Development in the Post-Cold War Era*. Geneva: UN Research Institute for Social Development.

Grosfoguel, Ramon and Ana Cervantes-Rodriguez (eds.). 2002. *The Modern/ Colonial Capitalist World-System in the Twentieth Century: Global Processes, Antisystemic Movements, and the Geopolitics of Knowledge*. Westport, CT: Greenwood.

Gulalp, Haldun. 1983. "Frank and Wallerstein Revisited: A Contribution to Brenner's Critique." In *Neo-Marxist Theories of Development*, ed. Peter Limqueco and Bruce McFarlane. New York: St. Martins.

Guthrie, Doug. 2006. *China and Globalization: the Social, Economic and Political Transformation of Chinese Society*. New York: Routledge.

Hagopian, Francis and Scott P. Mainwaring (eds.). 2005. *The Third Wave of Democratization in Latin America: Advances and Setbacks*. Cambridge, UK: Cambridge University Press.

Halebsky, Sander and Richard L. Harris (eds.). 1995. *Capital, Power, and Inequality in Latin America*. Boulder, CO: Westview.

Halevi, Joseph. 2002. "The Argentine Crisis." *Monthly Review* 53, 11 (April).

Halevi, Joseph and Bill Lucarelli. 2002. "Japan=s Stagnationist Crises." *Monthly Review* 53, 9 (February).

Hall, Burton (ed.). 1972. *Autocracy and Insurgency in Organized Labor*. New Brunswick, NJ: Transaction.

Halliday, Fred. 2001. *The World at 2000*. New York: St. Martins.

Halliday, John and Gavan Mc Cormic. 1973. *Japanese Imperialism Today*. New York: Monthly Review Press.

Hamilton, Clive. 1983. "Capitalist Industrialization in East Asia's Four Little Tigers." *Journal of Contemporary Asia* 13, 1.

———. 1986. *Capitalist Industrialization in Korea*. Boulder, CO: Westview.

Hamilton, Nora and Timothy F. Harding (eds.). 1986. *Modern Mexico: State, Economy, and Social Conflict*. Thousand Oaks, CA: Sage.

Hardt, Michael and Antonio Negri. 2000. *Empire*. Cambridge, MA: Harvard University Press.

Harris, Jerry. 2003. "Transnational Competition and the End of U.S. Hegemony." *Science and Society* 67, 1.

Harris, Richard (ed.). 1975. *The Political Economy of Africa*. Cambridge, MA: Schenkman.

Harrison, Bennett and Barry Bluestone. 1988. *The Great U-Turn: Corporate Restructuring and the Polarizing of America*. New York: Basic.

Hart, Jeffrey A. 1992. *Rival Capitalists: International Competitiveness in the United States, Japan, and Western Europe*. Ithaca, NY: Cornell University Press.

Hart-Landsberg, Martin. 1993. *Rush to Development: Economic Change and Political Struggle in South Korea*. New York: Monthly Review.

Hart-Landsberg, Martin and Paul Burkett. 2005. *China and Socialism: Market Reforms and Class Struggle*. New York: Monthly Review Press.

Harvey, David. 1982. *The Limits to Capital*. Chicago: University of Chicago Press.

———. 2003. *The New Imperialism*. New York: Oxford University Press.

———. 2006a. *A Brief History of Neoliberalism*. New York: Oxford University Press.

———. 2006b. *Spaces of Global Capitalism: A Theory of Uneven Geographical Development*. London: Verso.

Hassan, Salah S. and Erdener Kaynak (eds.). 1994. *Globalization of Consumer Markets: Structures and Strategies*. New York: International Business.

Haugerud, Angelique, M. Priscilla Stone, and Peter D. Little (eds.). 2000. *Commodities and Globalization*. Lanham, MD: Rowman and Littlefield.

Headley, Bernard. 1991. "The 'New World Order' and the Persian Gulf War." *Humanity and Society* 15, 3.

Hedley, R. Alan. 2002. *Running Out of Control: Dilemmas of Globalization*. West Hartford: Kumarian.

Held, David and Anthony Mc Grew. 2002. *Governing Globalization: Power, Authority and Global Governance*. Cambridge, UK: Polity.

———. 2002. *Globalization/Anti-globalization*. Cambridge, UK: Polity.

Held, David, Anthony G. McGrew, David Goldblatt, and Jonathan Perraton. 1999. *Global Transformations: Politics, Economics, and Culture*. Palo Alto, CA: Stanford University Press.

Held, David and Ayse Kaya (eds.). 2007. *Global Inequality: Patterns and Explanations*. Cambridge, UK: Polity.

Hertz, Noreena. 2002. *Silent Takeover: Global Capitalism and the Death of Democracy*. New York: The Free Press.

Higgott, Richard A. and Anthony Payne (eds.). 2000. *The New Political Economy of Globalisation*. Cheltenham, UK: Edward Elgar.

Hobson, John A. [1905] 1972. *Imperialism: A Study*. Rev. ed. Ann Arbor: University of Michigan Press.

Holloway, John. 1994. "Transnational Capital and the National State." *Capital & Class* 52.

Holzhausen, Arne (ed.). 2001. *Can Japan Globalize?: Studies on Japan=s Changing Political Economy and the Process of Globalization in Honour of Sung-Jo Park*. New York: Physica-Verlag.

Hoogvelt, Ankie M. M. 1982. *The Third World in Global Development*. London: Macmillan.

Hook, Glenn D. and Hasegawa Harukiyo (eds.). 2001. *Political Economy of Japanese Globalization*. New York: Routledge.

Hopkins, Terence K. and Immanuel Wallerstein. 1981. "Structural Transformations of the World-Economy." In *Dynamics of World Development*, ed. Richard Rubinson. Beverly Hills, CA: Sage.

Houtart, Francois and Francois Polet (eds.). 2001. *The Other Davos Summit: The Globalization of Resistance to the World Economic System*. London: Zed.

Howard, Andrew. 2005. "Global Capital and Labor Internationalism: Workers' Response to Global Capitalism." In *Globalization and Change: The Transformation of Global Capitalism*, ed. Berch Berberoglu. Lanham, MD: Lexington Books.

Howe, Carolyn. 1986. "The Politics of Class Compromise in an International Contex Considerations for a New Strategy for Labor." *Review of Radical Political Economics* 18, 3.

Howe, Irving (ed.). 1972. *The World of the Blue Collar Worker.* New York: Quadrangle.

Hsiung, Ping-Chun. 1995. *Living Rooms as Factories: Class, Gender, and the Satellite Factory System in Taiwan.* Philadelphia: Temple University Press.

Hudson, Michael. 2003. *Super Imperialism: The Origin and Fundamentals of U.S. World Dominance.* Rev. ed. Herndon, VA: Pluto.

Hudson, Yeager (ed.). 1999. *Globalism and the Obsolescence of the State.* Lewiston, NY: Edwin Mellen.

Hurrell, Andrew (ed.). 1999. *Inequality, Globalization, and World Politics.* New York: Oxford University Press.

Hutton, Will and Anthony Giddens (eds.). 2000. *Global Capitalism.* New York: The New Press.

Hytrek, Gary and Kristine M. Zentgraf 2008. *America Transformed: Globalization, Inequality and Power.* New York: Oxford University Press.

Institute for Labor Education and Research. 1982. *What's Wrong with the U.S. Economy?* Boston: South End.

James, Harold. 2002. *The End of Globalization: Lessons from the Great Depression.* Cambridge, MA: Harvard University Press.

Johnston, R. J., Peter J. Taylor, and Michael J. Watts. 2002. *Geographies of Global Change: Remapping the World.* Malden, MA: Blackwell.

Jones, R. J. Barry. 2000. *The World Turned upside Down? Globalization and the Future of the State.* Manchester: Manchester University Press.

Jung, Hwa Yol (ed.). 2002. *Comparative Political Culture in the Age of Globalization.* Lanham, MD: Rowman and Littlefield.

Kagarlitsky, Boris. 2000. *The Twilight of Globalization: Property, State, and Capitalism.* Herndon, VA: Pluto.

Kalb Don, Marco Van Der Land, Richard Staring, and Nico Wilterdink (eds.). 2000. *The Ends of Globalization: Bringing Society Back In.* Lanham, MD: Rowman and Littlefield.

Kang, David C. 2002. *Crony Capitalism: Corruption and Development in South Korea and the Philippines.* Cambridge: Cambridge University Press.

Kapstein, Ethan B. 1999. *Sharing the Wealth: Workers and the World Economy.* New York: Norton.

Karliner, Joshua. 1997. *The Corporate Planet: Ecology and Politics in the Age of Globalization.* Los Angeles: University of California Press.

Katsiaficas, George and Eddie Yuen (eds.). 2002. *The Battle of Seattle: Debating Capitalist Globalization and the WTO.* New York: Soft Skull.

Katz-Fishman, Walda, Jerome Scott, and Ife Modupe. 2005. "Global Capitalism, Class Struggle, and Social Transformation." In *Globalization and Change: The Transformation of Global Capitalism,* ed. Berch Berberoglu. Lanham, MD: Lexington.

Katznelson, Ira and Aristide R. Zolberg (eds.). 1986. *Working-Class Formation: Nineteenth-Century Patterns in Western Europe and the United States.* Princeton, NJ: Princeton University Press.

Kayizzi-Mugerwa, Steve. 2001. "Globalization, Growth and Income Inequality: The Africa Experience." OECD Development Center, Working Paper No 186.

Keeran, Roger. 1980. *The Communist Party and the Auto Workers' Unions.* Bloomington: Indiana University Press.

Kelly, Rita Mae, Jane H. Bayes, Mary Hawkesworth, and Brigitte Young (eds.). 2001. *Gender, Globalization, and Democratization.* Lanham, MD: Rowman and Littlefield.

Kennedy, Paul. 1987. *The Rise and Fall of the Great Powers.* New York: Random House.

Kennedy, Paul, Dirk Messner, and Franz Nuscheler (eds.). 2002. *Global Trends and GlobalGovernance.* Herndon, VA: Pluto.

Khor, Martin. 2001. *Rethinking Globalization.* London: Zed.

Kidron, Michael. 1970. *Western Capitalism since the War.* Rev. ed. Harmondsworth, UK: Penguin.

Kim, Samuel S. (ed.). 2000. *East Asia and Globalization.* Lanham, MD: Rowman and Littlefield.

Kim, Sueng-Kyung. 1997. *Class Struggle or Family Struggle?: The Lives of Women Factory Workers in South Korea.* Cambridge: Cambridge University Press.

Kimeldorf, Howard. 1988. *Reds or Rackets? The Making of Radical and Conservative Unions on the Waterfront.* Berkeley: University of California Press.

Kirkbride, Paul (eds.). 2001. *Globalization: The External Pressures.* Chichester: John Wiley and Sons, Ltd.

Klak, Thomas (ed.). 1997. *Globalization and Neoliberalism.* Lanham, MD: Rowman and Littlefield.

Kloby, Jerry. 1987. "The Growing Divide: Class Polarization in the 1980s." *Monthly Review* 39, 4 (September).

———. 1993. "Increasing Class Polarization in the United States: The Growth of Wealth and Income Inequality." In *Critical Perspectives in Sociology,* ed. Berch Berberoglu. Dubuque, IA: Kendall/Hunt.

Knapp, Peter and Alan J. Spector. 1991. *Crisis and Change: Basic Questions of Marxist Sociology.* Chicago: Nelson-Hall.

Kohl, Benjamin and Linda C. Farthing. 2006. *Impasse in Bolivia: Neoliberal Hegemony and Popular Resistance.* London: Zed.

Kozul-Wright, Richard and Robert Rowthorn (eds.). 1998. *Transnational Corporations and the Global Economy.* New York: St. Martins.

Lam Willy. 2009. "Hu Jintao's Great Leap Backward." *Far Eastern Economic Review* (January/February): 19–22.

Landau, Saul. 2003. *The Pre-Emptive Empire: A Guide to Bush's Kingdom.* London: Pluto.

Landsberg, Martin. 1979. "Export-Led Industrialization in the Third World: Manufacturing Imperialism." *Review of Radical Political Economics* 11, 4 (Winter).

———. 1988. "South Korea: The 'Miracle' Rejected." *Critical Sociology* 15, 3.

Larrain, Jorge. 1989. *Theories of Development: Capitalism, Colonialism, and Dependency.* Cambridge, UK: Polity.

Lechner, Frank J. and John Boli (eds.). 2007. *The Globalization Reader.* 3rd ed. Oxford: Wiley-Blackwell.

Lembcke, Jerry. 1988. *Capitalist Development and Class Capacities: Marxist Theory and Union Organization.* Westport, CT: Greenwood.

Lenin, V. I. [1917] 1975. *Imperialism: The Highest Stage of Capitalism.* In *Selected Works.* Vol. 1. Moscow: Foreign Languages Publishing House.

Levenstein, Harvey. 1981. *Communism, Anticommunism, and the CIO.* Westport, CT: Greenwood.

Levine, Marvin J. 1997. *Worker Rights and Labor Standards in Asia's Four New Tigers: A Comparative Perspective.* New York: Plenum.

Lewis, Cleona. 1938. *America's Stake in International Investments.* Washington, DC: Brookings.

Lim, Linda. 1983. "Capitalism, Imperialism, and Patriarchy: The Dilemma of Third World Women Workers in Multinational Factories." In *Women, Men, and the International Division of Labor,* ed. J. Nash and M. Fernandez-Kelly. Albany, NY: State University of New York Press.

Limqueco, Peter and Bruce McFarlane (eds.). 1983. *Neo-Marxist Theories of Development.* New York: St. Martin's.

Lofdahl, Corey L. 2002. *Environmental Impacts of Globalization and Trade.* Cambridge, MA: MIT Press.

Lotta, Raymond. 1984. *America in Decline.* Vol. 1. Chicago: Banner.

Luxemburg, Rosa. [1913] 1951. *The Accumulation of Capital.* Reprint. New Haven: Yale University Press.

Magbadelo, John Olushola. 2005. "Westernism, Americanism, Globalism and Africa's Marginality." *Journal of Third World Studies,* XXII, 2.

Magdoff, Harry. 1969. *The Age of Imperialism.* New York: Monthly Review.

———. 1978. *Imperialism: From the Colonial Age to the Present.* New York: Monthly Review.

———. 1992. *Globalization: To What End?* New York: Monthly Review.

———. 2003. *Imperialism without Colonies.* New York: Monthly Review Press.

Magdoff, Harry and Paul M. Sweezy. 1977. *The End of Prosperity: The American Economy in the 1970s.* New York: Monthly Review.

———. 1981. *The Deepening Crisis of U.S. Capitalism.* New York: Monthly Review.

———. 1987. *Stagnation and the Financial Explosion.* New York: Monthly Review.

Mahajan, Rahul. 2002. *The New Crusade: America=s War on Terrorism.* New York: Monthly Review.

Mahjoub, Azzam (ed.). 1990. *Adjustment or Delinking? The African Experience.* London: Zed.

Maitra, Priyatosh. 1996. *The Globalization of Capitalism in Third World Countries.* Westport, CT: Praeger.

Mandel, Ernest. 1975. *Late Capitalism.* London: New Left.

———. 1980. *The Second Slump.* London: Verso.

Mandel, Michael. 2004. *How America Gets Away with Murder.* Herndon, VA: Pluto.

Martin, Hans-Peter and Harold Schumann. 1997. *The Global Trap*. London: Zed.

Mayer, Tom. 1991. "Imperialism and the Gulf War." *Monthly Review* 42, 11.

Mayorga, Rene Antonio. 1978. "National-Popular State, State Capitalism and Military Dictatorship." *Latin American Perspectives* 5, 2.

McBride, Stephen and John Wiseman (eds.). 2000. *Globalization and Its Discontents*. New York: Macmillan.

McGrew, Anthony and David Held. 2007. *Globalization Theory: Approaches and Controversies*. Cambridge, UK: Polity.

McMichael, Philip. 2008. *Development and Social Change: A Global Perspective*. 4th ed. Thousand Oaks, CA: Pine Forge.

McMichael, Philip, James Petras, and Robert Rhodes. 1974. "Imperialism and the Contradictions of Development." *New Left Review* 85 (May–June).

McNally, David. 1991. "Beyond Nationalism, beyond Protectionism: Labor and the Canada-U.S. Free Trade Agreement." *Capital & Class*, 43 (Spring).

Meiksins Wood, Ellen, Peter Meiksins, and Michael Yates (eds.). 1998. *Rising from the Ashes?: Labor in the Age of "Global"Capitalism*. New York: Monthly Review.

———. 2005. *Empire of Capital*. London: Verso.

Mertes, Tom (ed.). 2004. *A Movement of Movements: Is Another World Really Possible?* New York: Verso, 2004.

Meyer, Mary K. and Elisabeth Prugl (eds.). 1999. *Gender Politics in Global Governance*. Lanham, MD: Rowman and Littlefield.

Milani, Brian. 2000. *Designing the Green Economy: The Post-industrial Alternative to Corporate Globalization*. Lanham, MD: Rowman and Littlefield.

Miles, Maria. 1990. *Patriarchy and Accumulation on a World Scale: Women in the International Division of Labor*. London: Zed.

Miller, John A. 1987. "Accumulation and State Intervention in the 1980s: A Crisis of Reproduction." In *The Imperiled Economy*. Bk. 1, ed. Robert Cherry, Christine D'Onofrio, Cigdem Kurdas, Thomas R. Michl, Fred Moseley, and Michele I. Naples. New York: Union for Radical Political Economics.

Mishel, Lawrence, Jared Bernstein and Heather Boushey. 2003. *The State of Working America, 2002–2003*. Ithaca, NY: Cornell University Press.

Mittleman, James H. 2000. *The Globalization Syndrome: Transformation and Resistance*. Princeton, NJ: Princeton University Press.

——— (ed.). 1997. *Globalization: Critical Reflections*. Boulder: Lynne Rienner.

Mittelman, James H. and Norani Othman (eds.). 2002. *Capturing Globalization*. New York: Routledge.

Moghadam, Valentine. 1993. *Democratic Reform and the Position of Women in Transitional Economies*. Oxford: Clarendon Press.

Mol, Arthur P. J. 2001. *Globalization and Environmental Reform*. Cambridge, MA: MIT Press.

Moody, Kim. 1988. *An Injury to All: The Decline of American Unionism*. London: Verso.

———. 1997. *Workers in a Lean World: Unions in the International Economy*. London: Verso.

Morris, George. 1971. *Rebellion in the Unions*. New York: New Outlook.

Morris, Nancy and Silvio Waisbord. 2001. *Media and Globalization: Why the State Matters*. Lanham, MD: Rowman and Littlefield.

Munck, Ronaldo. 2002. *Globalization and Labor: The New Great Transformation*. London: Zed.

Munck, Ronaldo and Peter Waterman (eds.). 1999. *Labour Worldwide in the Era of Globalization: Alternative Union Models in the New World Order*. New York: Palgrave Macmillan.

Murphy, Craig N. (ed.). 2002. *Egalitarian Politics in the Age of Globalization*. New York: Palgrave Macmillan.

Nabudere, Dan. 1977. *The Political Economy of Imperialism*. London: Zed.

———. 1981. *Imperialism in East Africa*. 2 Vols. London: Zed.

Naples, Nancy A. and Manisha Desai. 2007. *Women's Activism and Globalization*. New York: Pocket Books.

Nayar, Baldev Raj. 2001. *Globalization and Nationalism: The Changing Balance in India's Economic Policy, 1950–2000*. Thousand Oaks, CA: Sage.

Nissen, Bruce. 1981. "U.S. Workers and the U.S. Labor Movement." *Monthly Review* 33, 1 (May).

——— (ed.). 2002. *Unions in a Globalized Environment: Changing Borders, Organizational Boundaries, and Social Roles*. Armonk, NY: Sharpe.

Nonini, Donald M. 2008. "Is China becoming NeoLiberal?" *Critique of Anthropology* 28 (2): 145–176.

O'Connor, James. 1984. *Accumulation Crisis*. New York: Basil Blackwell.

Olson, W. 1985. "Crisis and Social Change in Mexico's Political Economy." *Latin American Perspectives* 46.

O, Meara, Patrick, Howard Mehlinger, and Matthew Krain (eds.). 2000. *Globalization and the Challenges of a New Century: A Reader*. Bloomington: Indiana University Press.

Ong, Aiwa. 1987. *Spirits of Resistance and Capitalist Discipline: Women Factory Workers in Malaysia*. Albany, NY: SUNY Press.

Owen, Roger and Bob Sutcliffe (eds.). 1972. *Studies in the Theory of Imperialism*. London: Longman.

Oxaal, Ivar, Tony Barnett, and David Booth (eds.). 1975. *Beyond the Sociology of Development*. London: Routledge & Kegan Paul.

Pakulski, Jan. 2004. *Globalizing Inequalities: New Patterns of Social Privelege and Disadvantage*. Crows Nest, Australia: Allen and Unwin.

Palast, Greg. 2002. *The Best Democracy Money Can Buy: An Investigative Reporter Exposes the Truth about Globalization, Corporate Cons, and High Finance Fraudsters*. London: Pluto.

Panitch, Leo and Colin Leys (eds.). 1999. *Global Capitalism versus Democracy*. New York: Monthly Review.

———. 2000. *Working Classes, Global Realities*. New York: Monthly Review.

———, 2003. *The New Imperial Challenge*. New York: Monthly Review.

Parenti, Michael. 1989. *The Sword and the Dollar: Imperialism, Revolution, and the Arms Race*. New York: St. Martin's.

———. 1995. *Against Empire*. San Francisco: City Lights.

Parrenas, Rhacel Salazar. 2001. *Servants of Globalization: Women, Migration and Domestic Work*. Stanford, CA: Stanford University Press.

Patnaik, Prabhat. 1999. "Capitalism in Asia at the End of the Millennium." *Monthly Review* 51, 3 (July/August).

Paxton, Pamela and Melanie M. Hughes. 2007. *Women, Politics, and Power: A Global Perspective.* Thousand Oaks, CA: Pine Forge.

Peet, Richard (ed.) 1987. *International Capitalism and Industrial Restructuring.* Boston: Allen & Unwin.

Perlo, Victor. 1988. *Super Profits and Crises: Modern U.S. Capitalism.* New York: International.

Perrucci, Carolyn C., Robert Perrucci, Dena B. Targ, and Harry R. Targ. 1988. *Plant Closings: International Context and Social Costs.* New York: Aldine de Gruyter.

Perrucci, Robert and Carolyn C. Perrucci. 2007. *The Transformation of Work in the New Economy: Sociological Readings.* Los Angeles: Roxbury.

Petras, James. 1978. *Critical Perspectives on Imperialism and Social Class in the Third World.* New York: Monthly Review Press.

———. 1981. *Class, State and Power in the Third World.* Montclair, NJ: Allanheld, Osmun.

———. 1997. "The Cultural Revolution in Historical Perspective." *Journal of Contemporary Asia* 27(4):445–459.

———. 1998. *The Left Strikes Back: Class Conflict in Latin America in the Age of Neoliberalism.* Boulder, CO: Westview.

———. 2002. "U.S. Offensive in Latin America: Coups, Retreats, and Radicalization." *Monthly Review* 54, 1 (May).

———. 2006. *The Power of Israel in the United States.* Atlanta: Clarity.

———. 2007. *Rulers and Ruled in the U.S. Empire.* Atlanta: Clarity.

———. 2008. *Zionism, Militarism, and the Decline of U.S. Power.* Atlanta: Clarity.

Petras, James and Christian Davenport. 1990. "The Changing Wealth of the U.S. Ruling Class." *Monthly Review* 42, 7 (December).

———. 1993. "Cultural Imperialism in the Late 20th Century." *Journal of Contemporary Asia* 23, 2.

Petras, James and Henry Veltmeyer. 2001. *Globalization Unmasked: Imperialism in the 21st Century.* London: Zed.

———. 2003. *System in Crisis: The Dynamics of Free Market Capitalism.* London: Zed.

———. 2007. *Multinationals on Trial: Foreign Investment Matters.* London: Ashgate.

Petras, James, Henry Veltmeyer, Luciano Vasapollo, and Mauro Casadio. 2006. *Empire With Imperialism: The Globalizing Dynamics of World Capitalism.* London: Zed.

Petras, James and Morris Morley. 1997. *U.S. Hegemony Under Siege: Class Politics and Development in Latin America.* London: Verso.

Philip, Gulley. 2003. *Democracy in Latin America: Surviving Conflict and Crisis?* Cambridge, UK: Polity.

Phillips, Brian. 1998. *Global Production and Domestic Decay: Plant Closings in the U.S.* New York: Garland.

Piazza, James A. 2002. *Going Global: Unions and Globalization in the United States, Sweden, and Germany.* Lanham, MD: Rowman and Littlefield.

Picciotto, Sol. 1990. "The Internationalization of the State." *Review of Radical Political Economics* 22, 1.

Pieterse, Jan Nederveen. 2007. *Globalization or Empire?* London: Taylor and Francis.

Polet, Francois (ed). 2007. *The State of Resistance: Popular Struggles in the Global South.* London: Zed.

Pollack, Mark A. and Gregory C. Shaffer (eds.). 2001. *Transatlantic Governance in the Global Economy.* Lanham, MD: Rowman and Littlefield.

Pollin, Robert. 2003. *Contours of Descent: U.S. Economic Fractures and the Landscape of Global Austerity.* London: Verso.

Prashad, Vijay and Teo Ballve (eds.). 2006. *Dispatches from Latin America: On the Frontlines Against Neoliberalism.* Boston: South End Press.

Prazniak Roxann and Arif Dirlik (eds.). 2001. *Places and Politics in an Age of Globalization.* Lanham, MD: Rowman and Littlefield.

Prugl, Elizabeth. 1999. *The Global Construction of Gender: Home-Based Work in the Political Economy of the 20th Century.* New York: Columbia University Press.

Rai, Shirin. 2001. *Gender and the Political Economy of Development: From Nationalism to Globalization.* Cambridge, UK: Polity.

Rajaee, Farhang. 2000. *Globalization on Trial: The Human Condition and the Information Civilization.* West Hartford, Conn.: Kumarian.

Rantanen, Terhi. 2002. *The Global and the National.* Lanham, MD: Rowman and Littlefield.

Rasler, Karen and William R. Thompson. 1994. *The Great Powers and Global Struggle, 1490–1990.* Lexington: University of Kentucky Press.

Reifer, Thomas E. (ed.). 2002. *Hegemony, Globalization and Anti-systemic Movements.* Westport, CT: Greenwood.

Rennstich, Joachim K. 2001. "The Future of Great Power Rivalries." In *New Theoretical Directions for the 21st Century World-System,* ed. Wilma Dunaway. Westport, CT: Greenwood.

Renton, Dave. 2002. *Marx on Globalization.* London: Lawrence & Wishart.

Ritzer, George. 2004. *The Globalization of Nothing.* Thousand Oaks, CA: Pine Forge.

Robbins, Richard H. 2002. *Global Problems and the Culture of Capitalism.* Boston: Allyn and Bacon.

Roberts, J. Timmons and Amy B. Hite. (eds.). 2007. *The Globalization and Development Reader:Perspectives on Development and Global Change.* Malden, MA: Blackwell Publishing.

———. 1999. *From Modernization to Globalization.* Oxford: Blackwell.

Robinson, William I. 1996. *Promoting Polyarchy: Globalization, U.S. Intervention, and Hegemony.* New York: Cambridge University Press.

———. 1998. "Beyond Nation-State Paradigms: Globalization, Sociology, and the Challenge of Transnational Studies." *Sociological Forum* 13.

———. 2004. *A Theory of Global Capitalism: Production, Class, and State in a Transnational World.* Baltimore, MD: The Johns Hopkins University Press.

Rosen, Ellen Israel. 2002. *Making Sweatshops: The Globalization of the U.S. Apparel Industry.* Los Angeles: University of California Press.

Ross, A. 1997. *No Sweat: Fashion, Free Trade, and the Rights of Garment Workers*. London: Verso.

Ross, Robert J. S.,1995. "The Theory of Global Capitalism: State Theory and Variants of Capitalism on a World Scale." In *A New World Order? Global Transformations in the Late Twentieth Century*, ed. David Smith and Jozsef Borocz. Westport, CT: Praeger.

Ross, Robert J. S. and Kent C. Trachte. 1990. *Global Capitalism: The New Leviathan*. Albany: SUNY Press.

Rowbotham, Sheila and Stephanie Linkogle (eds.). 2001. *Women Resist Globalization*. London: Zed.

Rowntree, Lester, Martin Lewis, Marie Price, and William Wyckoff. 2007. *Globalization and Diversity: Geography of a Changing World*, 2nd ed. Upper Saddle River, NJ: Prentice Hall.

Roy, Ash Narain. 1999. *The Third World in the Age of Globalization*. London: Zed.

Saenz, Mario (ed.). 2002. *Latin American Perspectives on Globalization*. Lanham, MD: Rowman and Littlefield.

Sassen, Saskia. 1998. *Globalization and Its Discontents*. New York: The New Press.

———. 2009. "Too Big To Save: The End of Financial Capitalism." *Open Democracy News Analysis*, April 2.

Schaeffer, Robert K. 2003. *Understanding Globalization: The Social Consequences of Political, Economic, and Environmental Change*. 2nd ed. Lanham, MD: Rowman and Littlefield.

Scholte, Jan Aart. 2000. *Globalization: A Critical Introduction*. New York: Macmillan.

Schoonmaker, Sara. 2002. *High-Tech Trade Wars: U.S.-Brazilian Conflicts in the Global Economy*. Pittsburgh, PA: University of Pittsburgh Press.

Servan-Schreiber, J. J. 1968. *The American Challenge*. New York: Atheneum.

Shepard, Benjamin and Ronald Hayduk (eds.). 2002. *From ACT UP to the WTO: Urban Protest and Community Building in the Era of Globalization*. London: Verso.

Sherman, Howard. 1976. *Stagflation*. New York: Harper & Row.

———. 1987. *Foundations of Radical Political Economy*. New York: Sharpe.

Siebert, Horst. 2000. *Globalization and Labor*. Kiel, Germany: Kiel Institute of World Economics.

Simon, Rick. 2000. "Class Struggle and Revolution in Eastern Europe: The Case of Poland." In *Marxism, the Millennium, and Beyond*, ed. Mark Cowling and Paul Reynolds. New York: Macmillan.

Siochru, Sean O. and W. Bruce Girard with Amy Mahan. 2002. *Global Media Governance*. Lanham, MD: Rowman and Littlefield.

Sklair, Leslie. 1989. *Assembly for Development: The Maquila Industry in Mexico and the United States*. Boston: Unwin Hyman.

———. 1991. *Sociology of the Global System*. Baltimore: Johns Hopkins University Press.

———. 1999. *Sociology of the Global System*. Baltimore: The Johns Hopkins University Press.

———. 2001. *The Transnational Capitalist Class*. Malden, MA: Blackwell.

———. 2002. *Globalization: Capitalism and Its Alternatives*. New York: Oxford University Press.

———. 2007. *The Sociology of Progress*. London: Routledge.

Smith, Jackie G. and Hank Johnston (eds.). 2002. *Globalization and Resistance: Transnational Dimensions of Social Movements*. New York: Routledge.

So, Alvin Y. 2003. "The Making of a Cadre-Capitalist Class in China." In *China's Challenges in the Twenty-First Century*, ed. Joseph Cheng. Hong Kong: City University of Hong Kong Press.

———. 2005. "Beyond the Logic of Capital and the Polarization Model: The State, Market Reforms, and the Plurality of Class Conflict in China." *Critical Asian Studies* 37(3): 481–494.

———. 2007. "The State and Labor Insurgency in Post-Socialist China." In *Challenges and Policy Programs of China s New Leadership*, ed. Joseph Cheng. Hong Kong: City University of Hong Kong Press.

———. 2008. "Peasant Conflict and the Local Predatory State in the Chinese Countryside." *Journal of Peasant Studies* 34(3–4): 560–581.

Stalker, Peter. 2000. *Workers without Frontiers: The Impact of Globalization on International Migration*. Boulder, CO: Rienner.

Starr, Amory. 2001. *Naming the Enemy: Anti-corporate Movements Confront Globalization*. London: Zed.

Stephens, Philip. 2009. "A Summit Success That Reflects a Different Global Landscape." *Financial Times*, April 3: 9.

Steven, Rob. 1994. "New World Order: A New Imperialism." *Journal of Contemporary Asia* 24, 3.

Stevis, Dimitris and Terry Boswell. 2008. *Globalization and Labor: Democratizing Global Governance*. Lanham, MD: Rowman and Littlefield.

Stiglitz, Joseph E. 2007. *Making Globalization Work*. New York: Norton.

———. 2002. *Globalization and Its Discontents*. New York: Norton.

Sugihara, Kaoru. 1993. "Japan, the Middle East and the World Economy." In *Japan in the Contemporary Middle East*, ed. K. Sugihara and J. A. Allan. London: Routledge.

Sutcliff, R. and Roger Owen (eds.). 1972. *Studies in the Theory of Imperialism*. London: Longman.

Suter, Keith. 2000. *In Defense of Globalization*. Sydney: New South Wales University Press.

———. 2002. *Global Order and Global Disorder: Globalization and the Nation-State*. Westport, CT: Praeger.

Sweezy, Paul M. and Harry Magdoff. 1972. *The Dynamics of U.S. Capitalism*. New York: Monthly Review.

———. 1988. "The Stock Market Crash and Its Aftermath." *Monthly Review* 39, 10 (March).

Szymanski, Albert. 1974. "Marxist Theory and International Capital Flows." *Review of Radical Political Economics* 6, 3 (Fall).

———. 1978. *The Capitalist State and the Politics of Class*. Cambridge, MA: Winthrop.

Szymanski, Albert. 1981. *The Logic of Imperialism.* New York: Praeger.

Tabb, William K. 1997. AGlobalization Is *An* Issue, the Power of Capital Is *the* Issue, *Monthly Review* 49, 2 (June).

———. 2001. *The Amoral Elephant: Globalization and the Struggle for Social Justice in the Twenty-first Century.* New York: Monthly Review.

Tanzer, Michael. 1974. *The Energy Crisis: World Struggle for Power and Wealth.* New York: Monthly Review.

———. 1991. "Oil and the Gulf Crisis." *Monthly Review* 42, 11.

Teeple, Gary. 1995. *Globalization and the Decline of Social Reform.* Amherst, NY: Humanity.

The World Bank. 2007. *The Global Citizen's Handbook: Facing Our World's Crises and Challenges.* New York: Harper Collins.

Thomas, Caroline and Peter Wilkin (eds.). 1997. *Globalization and the South.* New York: St. Martin's Press.

Thomas, Janet. 2000. *The Battle in Seattle: The Story behind the WTO Demonstrations.* Golden, CO: Fulcrum.

Thompson, William R. 2000. *The Emergence of the Global Political Economy.* New York: Routledge.

Tonelson, Alan. 2000. *The Race to the Bottom: Why a Worldwide Worker Surplus and Uncontrolled Free Trade are Sinking American Living Standards.* Boulder, CO: Westview.

Turner, Lowell, Harry C. Katz, and Richard W. Hurd (eds.). 2001. *Rekindling the Movement: Labor's Quest for Relevance in the Twenty-First Century.* Ithaca, NY: ILR Press.

Valvano, Vince. 1988. "No Longer #1? Assessing U.S. Economic Decline." *Dollars & Sense,* 142 (December).

Vanden, Harry E. and Gary Prevost 2006. *Politics of Latin America: The Power Game.* New York: Oxford University Press.

Van der Pijl, Kees. 1998. *Transnational Classes and International Relations.* New York: Routledge.

Vayrynen, Raimo (ed.). 1999. *Globalization and Global Governance.* Lanham, MD: Rowman and Littlefield.

Vellinga, Menno (ed.). 1999. *Dialectics of Globalization: Regional Responses to World Economic Processes: Asia, Europe, and Latin America in Comparative Perspective.* Boulder, CO: Westview.

Veltmeyer, Henry. 1999. "Labor and the World Economy." *Canadian Journal of Development Studies* 20 (Special Issue).

——— (ed.). 2008. *New Perspectives on Globalization and Antiglobalization: Prospects for a New World Order?* Burlington, VT: Ashgate.

Vietor, Richard. H. K. and Robert E. Kennedy. 2001. *Globalization and Growth: Cases in National Economics.* Fort Worth, TX: Harcourt.

Wachtel, Howard M. 1986. *The Money Mandarins: The Making of a Supranational Economic Order.* New York: Pantheon.

Wagner, Helmut (ed.). 2000. *Globalization and Unemployment.* New York: Springer.

Wallach, Lori and Michelle Sforza. 1999. *Whose Trade Organization?: Corporate Globalization and the Erosion of Democracy*. Washington, DC: Public Citizen.

———. 2000. *The WTO: Five Years of Reasons to Resist Corporate Globalization*. New York: Seven Stories.

Wallerstein, Immanuel. 1974. "The Rise and Future Demise of the World Capitalist System." *Comparative Studies in Society and History* 16, 4 (September).

——— 1979. *The Capitalist World-Economy*. Cambridge: Cambridge University Press.

———. 2002. "The United States in Decline?" In *Hegemony, Globalization, and Anti-Systemic Movements*, ed. Thomas E. Reifer. Westport, CT: Greenwood.

———. 2003. *The Decline of American Power: The United States in a Chaotic World*. New York: New Press.

Warren, Bill. 1973. "Imperialism and Capitalist Industrialization." *New Left Review* 81 (September/October).

———. 1980. *Imperialism, Pioneer of Capitalism*. New York: Verso.

Waterman, Peter. 1998. *Globalization, Social Movements, and the New Internationalisms*. London: Mansell.

Waters. Malcolm. 1995. *Globalization: The Reader*. New York: Routledge.

Webber, Michael John. 1996. *The Golden Age Illusion: Rethinking Postwar Capitalism*. New York: Guilford.

Weber, Steven (ed.). 2001. *Globalization and the European Political Economy*. New York: Columbia University Press.

Weeks, John. 1981. *Capital and Exploitation*. Princeton, NJ: Princeton University Press.

Went, Robert. 2000. *Globalization: Neoliberal Challenge, Radical Responses*. Trans. Peter Drucker. London: Pluto.

Worrell, Rodney. 2001. "Whither Global Africa? A Case of Pan-Africanism." *Africa Quarterly*, 41, 1–2.

Wright, Erik Olin. 2005. *Approaches to Class Analysis*. Cambridge: Cambridge University Press.

Yang, Xiaohua. 1995. *Globalization of the Automobile Industry: The United States, Japan, and the Peoples' Republic of China*. Westport, CT: Praeger.

Yates, Michael D. 2003. *Naming the System: Inequality and Work in the Global Economy*. New York: Monthly Review.

Yuen, Eddie, George Katsiaficas, and Daniel Burton Rose (eds.). 2002. *The Battle of Seattle: The New Challenge to Capitalist Globalization*. New York: Soft Skull.

Zloch-Christy, Iliana (ed.). 1998. *Eastern Europe and the World Economy: Challenges of Transition and Globalization*. Cheltenham, UK: Edward Elgar.

Zupnick, Elliot. 1999. *Visions and Revisions: The United States in the Global Economy*. Boulder, CO: Westview.

Contributors

Berch Berberoglu is Professor and Chair of the Department of Sociology and Director of Graduate Studies in Sociology at the University of Nevada, Reno. He received his Ph.D. in sociology from the University of Oregon in 1977. His areas of specialization include globalization, class analysis, political economy of development, and Third World studies. He is the author and editor of twenty-seven books and many articles, including *Labor and Capital in the Age of Globalization* (Rowman and Littlefield, 2002), *Globalization of Capital and the Nation-State* (Rowman and Littlefield, 2003), *Globalization and Change* (Lexington Books, 2005), and *Class and Class Conflict in the Age of Globalization* (Lexington Books, 2009).

Lourdes Benería is Professor in the Department of City and Regional Planning and the program on Gender, Feminist and Sexuality Studies at Cornell University. She has published numerous articles on women and work, the informal economy, globalization, gender and development, and other topics. Among her most recent books, she is the author of *Gender, Development and Globalization* (Routledge, 2003) and the editor of *Global Tensions: Challenges and Opportunities in the World Economy* (Routledge, 2003), with Savitri Bisnath, and of *Rethinking Informalization: Poverty, Precarious Jobs and Social Protection* (Cornell e-publishing, 2006), with Neema Kudva.

Johnson W. Makoba is Associate Professor of Sociology in the Department of Sociology at the University of Nevada, Reno. He received his Ph.D. in sociology from the University of California, Berkeley in 1990. His areas of specialization are development, organizations and bureaucracies, Third World studies, and Africa. He is the author of *Government Policy and Public Enterprise Performance in Sub-Saharan Africa* (Edwin Mellen Press, 1998). He is currently working on a book on the role of nongovernmental organizations (NGOs) in the development process in East Africa.

Martin Orr is Associate Professor and Chair of the Department of Sociology at Boise State University in Boise, Idaho. He received his Ph.D. in sociology from the University of Oregon in 1992. His research and teaching interests include social inequality, political sociology, globalization, social movements, and the media. He is the author of numerous articles on issues related to globalization, social inequality, social movements, and social change.

James Petras is Emeritus Professor of Sociology, State University of New York at Binghamton and Adjunct Professor at St. Mary's University in Halifax, Nova Scotia, Canada. He received his Ph.D. from the University of California at Berkeley in political science in 1967. His areas of specialization include the political economy of the U.S. Empire, social movements, and Latin America. He has published sixty-three books in twenty-nine languages and over 500 articles. He is the coauthor (with Henry Veltmeyer) of *Globalization Unmasked* (Zed Books, 2001), *System in Crisis* (Zed Books, 2003), and *What is Left in Latin America* (Ashgate, 2009).

Jan Nederveen Pieterse is Mellichamp Professor of Global Studies and Sociology in the Global and International Studies Program at the University of California, Santa Barbara. He received his Ph.D. in social science at the University of Nijmegen, Netherlands in 1988. His areas of specialization include globalization, political economy, development and cultural studies. He is the author and editor of fifteen books and many articles, including *Global Futures: Shaping Globalization* (Zed Books, 2000), *Globalization or Empire?* (Routledge, 2004), *Politics of Globalization* (Co-edited, Sage, 2009), *Is There Hope for Uncle Sam? Beyond the American Bubble* (Zed Books, 2008), and *Globalization and Emerging Societies* (Co-edited, Palgrave, 2009).

Alvin Y. So is Professor in the Division of Social Science at Hong Kong University of Science and Technology in Hong Kong, China. He received his Ph.D. in sociology from the University of California, Los Angeles in 1982. His research interests include social classes, development, and China. He is the editor of *China's Developmental Miracle: Origins, Transformations, and Challenges* (East Gate Book, 2003) and editor (with Mark Selden) of *War and State Terrorism: The U.S., Japan, and Asia-Pacific in the Long Twentieth Century* (Rowman and Littlefield, 2004).

Alan J. Spector is Professor of Sociology at Purdue University Calumet in Hammond, Indiana. He received his Ph.D. in sociology from Northwestern University in Evanston, Illinois in 1980. His areas of teaching and research interests include political economy, globalization, social movements, and social change. He is currently on the editorial board of the journal *Critical Sociology* and is the coauthor (with Peter Knapp) of *Crisis and Change: Basic Questions of Marxist Sociology* (Nelson-Hall, 1991).

Henry Veltmeyer is Professor of Development Studies at the Universidad Autónoma de Zacatecas (UAZ) in Mexico and Professor of Sociology and International Development Studies at St. Mary's University in Halifax, Nova Scotia, Canada. He received his Ph.D. in Political Science from MacMaster University in Hamilton, Ontario, Canada in 1976. He specializes in Latin America, development theory, and the political economy of development. He is author and editor of some thirty books, including (with James Petras) *Globalization Unmasked* (Zed Books, 2001), *System in Crisis* (Zed Books, 2003), and *What's Left in Latin America* (Ashgate, 2009).

Index

Printed in the United States
By Bookmasters